MERCEDES - BENZ

The Mercedes SLK R171

From the SLK200 Kompressor

to the

SLK55 AMG Black Series

(2004 – 2011)

By Bernd S. Koehling

ISBN: 979-8736812097

CONTENT

Technical Chapters

FOREWORD

First, I would like to thank you for having purchased this book. I hope you will enjoy reading it as much as I enjoyed writing it. This revised book covers the SLK R171, built from 2004 until 2011 with all its variations, AMG version, special editions and some of the tuner activities such as models from Brabus, Piecha and Renntech. Although there were a lot more companies around that dealt in one way or another with its tuning, it would have been impossible to cover them all, so I ask you for forgiveness, that the list that made it into this book is somewhat subjective.

When the first generation SLK was shown in 1994 at the Turin auto show as a concept car, it was a surprise to everyone, who was a bit familiar with Daimler-Benz. For the company it was not just a new car in a new segment, it was a quantum leap forward and put the company at once ahead of its archrivals Porsche and BMW, which had similar sized cars on offer. One of its hallmarks was of course its automatic vario-roof, which proved so popular that it helped to propel the new car ahead of its competition.

By 2004, its novelty had worn off and a replacement needed to be introduced. Gone were the days of elegant styling that lacked a bit true sports car flair. The second generation looked with its Formula One/SLR derived prominent front much more masculine. This was supported by the fact that the car came only with one four-cylinder engine. A smooth and powerful 3.5 L V6 was supposed to be the main source of energy, yet it came even better, when pretty much from the beginning also an AMG version was available with a first in this class of cars: a phenomenal 5.4 L naturally aspirated V8.

Daimler-Benz was off to a good start with its latest small sports car creation and occasional problems with its V6 balance shaft sprockets could not stop it from becoming another sales success.

I would like to thank the members of benzworld.org, and especially slkworld.com for the invaluable information I was able to gather from their input. With the publication of this hardcover edition, some of the technical information has been updated and more photos added.

April 2021

Bernd S. Koehling

Mercedes-Benz SLK R171 series
2004 - 2011

How it all began

Whoever wants to know more about the SLK, will at one point also like to know more about the vehicle concept`s history. After all, a name similar to the SLK name was synonymous way back in the 1920s and 1930s with racing successes across Europe. The cars that carried that name (or part of it) were instrumental in establishing the Mercedes brand firmly as a major entry in the sports car market. Of course the huge and powerful SS (Super, Sport), SSK (Super, Sport, Kurz or Kompact) and SSKL (Super, Sport, Kurz or Kompact, Light) of those days have nothing really in common with their "tiny" modern cousin.

Still, even a distant relationship through a loose similarity in the name does never hurt. In order to find other traces within the long and esteemed Daimler-Benz history, one has to go to the 1950s. A certain Max Hoffman had pushed the executive board to create a special automobile that would help

him sell Mercedes cars in the New World. It would also be responsible to make the brand for the first time known to a larger circle in the US; people, who had previously heard very little about a company called Daimler-Benz. And if they did, it was not necessarily positive. The car is of course the 300SL Gullwing, shown first to the public at the New York automobile show in 1954.

But the Daimler-Benz management discussed with Hoffman already prior to the launch of the iconic SL that a smaller, more affordable roadster would be a beneficial addition to such an important market as North America.

After one had finally agreed on the design, it took Daimler-Benz stylists just eight weeks from blueprints to a first 1:1 scale model. The car was of course the 190SL. And it stood side by side with the 300SL at said show in 1954.

For both, Daimler-Benz and Hoffman, the decision to offer a smaller SL was purely market driven. Everybody knew that the ultra-expensive 300SL would not sell in large numbers. But everybody also knew that a more affordable, similarly styled roadster would attract a much bigger crowd that was intrigued by the aura of the Über-SL, but not necessarily by the price tag that came with it.

Had it been only for the 300SL, the SL sports car would not have seen a successor.

After all, both coupe and roadster sold a mere 3,258 units combined (coupe: 1,400, roadster: 1,858) from 1954 to 1963. That means on average less than 30 cars a month. The "lesser" SL managed to sell between 1955 and 1963 a respectable 25,881 units and convinced the executive board to go ahead with the development of a new version of the SL. As we all know, it was the 230SL pagoda, launched in 1963.

A 1929 Model SSK

The prototype 190SL in 1954 in New York

Fast forward to 1989:

After the long and successful career of the pagoda successor R107, the R129 was launched in 1989. Initial production was planned for 20,000 units annually. Although prices for the new roadster had reached almost stratospheric levels, demand far outweighed supply. At the top end of the convertible car market, the SL had in its price range no rivals. Capacity at the newly opened Bremen plant was soon increased to 25,000 units annually, but to no avail. The waiting list hovered around two years and could even reach up to five years for models in unique equipment options. The situation was similar to the 1970s, where there was a famous saying among German farmers, who had ordered a Mercedes Diesel sedan: "I can cope with draughts and floods, but not with the long waiting list for my new diesel".

And then suddenly in 1990 everything changed: the Mazda Miata hit earth. Although it was not really a threat to the upper end of the convertible car market, Daimler-Benz management knew instantly: this was a game changer. Of course there were other convertibles available at that time. The Cadillac Allante, produced from 1987 to 1993, was an attempt by GM to steal some of the SL`s glamour. Although it was a bargain compared with the SL and although it offered a state-of-the-art Northstar DOHC V8, people could not warm to it in sufficient numbers..

Mazda Miata (MX5) from 1991 and Cadillac Allante (next page) from 1992

12

And flying it during its production process from Italy (Pininfarina had designed it and built the body) to the US did not really help to improve bottom line. So when sales failed to impress management, the project was abandoned. Overall just 21,430 Allantes were built. Cadillac tried the convertible market again only in 2003 with the XLR. But this time with a retractable metal hardtop, which was designed and built by a joint venture company between Mercedes and Porsche, which also supplied the SLK's vario roof.

The lovely Ford Mustang (especially with its 5.0 l engine) was a joy to drive, but was not offered by Ford outside the US. Much cheaper than the SL, it was also not meant to be a match for the SL`s built quality. And the BMW Z1 with its troublesome disappearing doors (and lots of interesting

technical features for its time) was offered as a limited edition model with 8,000 units sold from 1989 to mid 1991.

The Miata was different. It would make people suddenly realize that open top driving was within their financial reach. The small car was not only attractively priced and soundly engineered; it was on top of that gorgeous to look at and managed to hit the emotional soft spot in most drivers, males and females. As a consequence of all this, it was THE car to be seen in. This included "the rich and famous". Their today`s Prius was the Miata in the early 1990s. What the British had achieved so successfully in the years after WWII with cars such as the brilliant MG TC and later the MGA, the Japanese had now simply copied.

And it is somewhat ironic that a team from Mazda had spent quite some time on the stand of the British "Stevens Cipher" roadster at the 1980 Birmingham Motor Show in the UK. Their later mission statement for the Miata is supposed to be a straight copy of the press release for the Cipher (according to Prof. Tony Stevens, Chairman of Stevens Research Ltd).

Daimler-Benz (together with other car companies) was keenly aware that this "small convertible niche" would not just grow the way the overall automotive market would expand; no, they realized that it would grow by leaps and bounds. And this would not happen only in the US and in Europe, it would happen all over the world. They urgently needed some presence there. Not at the price level of the Mazda. That was not their domain (yet), but size-wise their new Mercedes convertible had to be much smaller (and more affordable) than the R129.

It was rather fitting that the company had launched, after a long absence of some fifty years, a mid-prized sedan in form of the 190 W201 in 1982 again. A small and affordable roadster would make a more than welcome addition to this new line of cars and it was only natural that parts from the new W202 C-class were shared with the R170 sports car. After all it was just the same what Daimler-Benz had already practiced in the 1950s with the 180/190 sedan and the 190SL.

The new baby-SL could not just be a slightly larger Miata on steroids, which Daimler-Benz could sell at a premium. They knew the new car had to come with a twist, a USP that nobody would have expected from this traditional minded company. Luckily the days were long gone, when Daimler-Benz was regarded by journalists as an automotive manufacturer that would only produce high quality cars for executives, farmers and well-healed retirees.

Also the second generation SLK has not lost any of the appeal of its predecessor

14

History of the vario roof

Just in case the esteemed reader is interested in the history of the vario-roof prior to the first generation SLK, then this chapter is for you. Otherwise please proceed to the SLK vario-roof chapter.

In most publications we will be informed that the first car equipped with a vario-roof was a Peugeot 401 from 1934. This is correct, when we consider electrical hardtops only. But that is just half the story, as the concept to have an all-weather removable metal top in an automobile dates further back. And it did not occur in France, but in the US. Benjamin Ellerbeck from Salt Lake City, Utah was an engineer with a keen interest in everything automotive. By autumn of 1919 he had worked out the basic design of a shiftable/ retractable top for open cars. By the early 1920s he had built a number of 1/8 scale models to demonstrate potential clients, how his invention worked. In 1921 he was granted a patent (US patent no. 1.379.906) for his shiftable top. Jumping a bit forward here, in December 1930 he was granted another US patent for a roadster that featured a second windshield for the rumble seat.

Although he was fairly enthusiastic about his invention, he had so far failed to attract a major customer to apply his ideas. So in 1922 Ellerbeck bought a 1919 Hudson Super Six, which should serve as real world model for his innovative top. In order to make his new roof fit, he had to rebuild major parts of the Hudson. But the result was an attractive car, which looked especially inviting with the top down. Once closed, the overhang over the windshield looked similar to a giant air scoop or sun visor.

One problem was the fairly complicated mechanism to manually raise and lower the top. Costly to produce, it scared off quite a few possible customers. So in 1923 he changed his creation by letting the roof rest on landau bars, which were secured to the ends of a cross shaft passing through the body. The previous creation offered straight diagonal arms, which were located inside of both body and top. The top was a metal frame with either fabric or metal cover. In lowered position it would settle flush on the rear deck. In either lowered or raised position it was secured with clamps.

Still, despite his design changes, he had not found an interested party for the innovative top. So in 1925 he wrote a lengthy letter to Packard Motor Company, trying to interest them in his creation. In his letter he stated that he understood the car manufacturer's desire to draw the line somewhere in the diversity of body styles. Then he continued in his letter that he was convinced that a Packard roadster would leap in favor if given a modern top construction. Unfortunately, Packard declined, but this did not stop his enthusiasm for his design.

Ellerbeck even managed to get his design covered in an article by a British car magazine in the early 1930s, but also saw no interest from a foreign manufacturer forthcoming. The 1929 stock market crash and following economic crisis certainly did not help and when the interest in roadsters decreased towards the mid-1930s, so did further activities from Benjamin Ellerbeck, who had fought so hard for over 15 years to have his innovative concept accepted. Like so many good ideas, it was probably ahead of its time.

The top for the Hudson was altered, as it now sat on top of the car's rear part

An opening in the roof left room for additional passengers

Although almost forgotten, the concept was not dead entirely. Ellerbeck's idea was picked up in 1930 in Europe by a French dentist, who had just like Ellerbeck a keen interest in everything automotive. Because next to making a living as a dentist, George Paulin was also a gifted part-time automobile designer. The shiftable top concept was re-evaluated, simplified and it finally evolved under Paulin into the first power-operated retractable hardtop and was consequently patented by him in 1931.

But, as fate had it also on this side of the Atlantic, no one was interested in his concept. That changed luckily in autumn 1933, when Peugeot's Paris car dealer Emile Darl`mat (who was a friend of Paulin), introduced the dentist to French coach-builder Marcel Pourtout. Pourtout was immediately intrigued by Paulin's patent and all three worked on making the retractable top happen. It was Paulin's luck to have a large car dealer and a respected coach-builder at his side, a support that Ellerbeck unfortunately never enjoyed.

In May 1934, Carosserie Pourtout used a mid-sized Peugeot 402BL, supplied by Darl`mat, to introduce its Eclipse Decapotable (retractable roof). The revolutionary car gained of course quite some news coverage, which in turn brought Peugeot management into contact with Paulin. In 1935 he convinced the company of the virtues of his novel concept and consequently sold them his patent.

In order to further support the idea, Paulin worked from 1934 to 1938 as Pourtout's designer and helped to launch the system also on cars such as the Peugeot 301, 401 and 601. In total 79 Peugeot Type 401 and 473 Peugeot Type 402 were produced. Sufficient space for the large metal roof was vital, so 402 cars of all vehicles produced were built on the extended sedan chassis of the "Familiale Limousine", which had a length of some 5,30 m (210.4 in). The extended chassis made the cars relatively expensive.

So in order to save on costs, it was decided to offer those cars only with a manually operated top. Most customers did not mind, as it was luckily relatively easy to manually raise or lower the top. Today only 34 of the Eclipse Decapotable have survived. Other vehicles designed with retractable roof by Paulin and produced by Pourtout included the Italian Lancia Belna and various models from French car manufacturers Hotchkiss and Panhard.

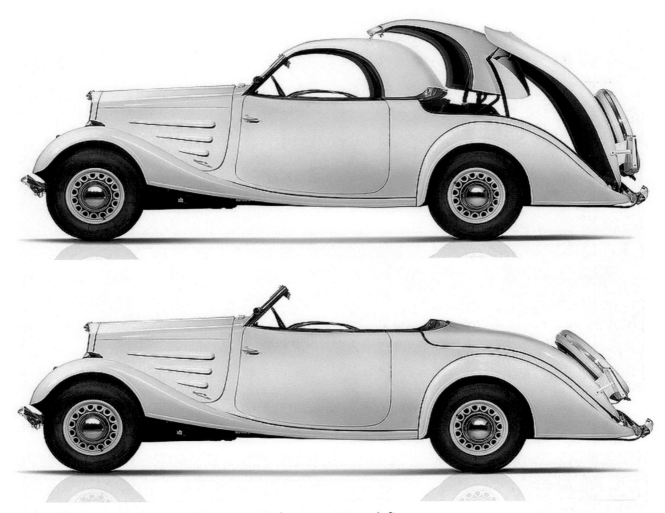

Once the top was lowered, there was no luggage space left

That would have been the end of the retractable hardtop, had Ford not picked up the idea again in 1956 to offer a Fairlane 500 Skyliner for the model years 1957 to 1959. Part of the Fairlane model line-up, the Skyliner was a full-sized two-door offer with a fairly complex electrically operated hardtop. The top`s front folded and disappeared with the rest of the roof under a long trunk lid. Compared with the Peugeot, this was high-tech pure. The driver did not have to do anything but to push a single button and everything else happened by miracle. The system was equipped with three electrical motors that drove four jacks to lift and stow the top, four locking mechanisms to hold the top in place, ten solenoids and another four electrical motors to lock the doors and trunk. Each of the seven electrical motors had its own circuit breaker and 186 m (610 ft) of wiring was required to make this all work.

It looked complicated, but was reliable and must have been stunning in the 1950s

With a length of 5.35 m (210.8 in), the big Ford was just 5 mm longer than its French predecessor, which did not have the split roof. And according to period reviews the Ford's top worked despite its complicated set-up surprisingly reliable. Ford produced in those three years a total of 38,394 units.

While we are in the US, there was one more US car that used a retractable hardtop. It was created much earlier than the Ford and was the 1941 Chrysler Thunderbolt Roadster, which was produced to help improve Chrysler's image among consumers. It must have been one of the first American dream cars and was designed by

Alex Tremulis. Four of the five Thunderbolt ever built are known to have survived.

Mitsubishi offered in 1995 and 1996 a special edition of its 3000GT model. It was shipped from its Nagoya, Japan factory to ASC (American Sunroof Company) in California in coupe form. There it was transformed to the first modern vario-roof car. Mitsubishi marketed it as "3000GT Spyder". The roof opened fully automatic with the push of a button. But with a hefty price tag of $65,000 only 1,618 units could be sold, so it was abandoned in 1996. The car must have been Mitsubishi's quick answer to the 1994 SLK presentation in Turin and Paris.

The 1941 Chrysler Thunderbolt Roadster. Very little is known about this car. Its turning radius must have been huge

The Mitsubishi 3000GT Spyder

The SLK vario-roof

The basic vario-roof system was unchanged, when compared with the R170. It was still basically divided in two halves, which were linked by a kinematic mechanism that was automatically locked when the roof was closed. A big change was the disconnection of the rear window from the C-pillars. In order to increase luggage space, the rear window rotated now by 150 degrees when the roof was stored in the trunk.

At the touch of a button on the right side of the centre console, a hydraulic system with five cylinders controlled the folding process. This included the opening and closing of the trunk lid, which opened by tipping to the rear. This was necessary, because that way the roof halves had enough space to pivot backwards. But this process also meant that a minimum clearance height of 1.65 m (5.5 ft) and an additional 0.25 m (10 in) length behind the SLK was required to lower or raise the hardtop.

The roof's front latch hydraulic cylinder operated two latches at the front of the roof. Two hydraulic cylinders stowed the roof panel, rear window and C-pillars in the upper section of the trunk. Two further hydraulic cylinders operated the trunk lid. As with the R170, a plastic cover (which was now a solid plastic tray) separated it from the luggage space below to prevent luggage items from colliding with the roof. This luggage cover could be pivoted longitudinally and the vario-roof could only be opened if the cover was closed.

Once the roof was closed, the cover could be easily moved forward by hand, which in turn helped loading and unloading. Top down, the trunk offered a capacity of 208 l (7.3 cu ft), which was some 63 l more than the R170 offered. This should be sufficient for two not too large sport bags. With the top out of the way, the load volume increased to 300 l (10.6 cu ft), which was a bit surprisingly 48 l less than the predecessor had offered. Now also high drink cases could be stowed. The driver could operate the hardtop only, when the car was not traveling faster than 8 km/h (5 mph).

The roof created quite a stir in the early SLK days; it was further refined with the launch of the R171

For the first time, Daimler-Benz offered an infrared remote roof control integrated in the standard SmartKey, when the Premium package (North America) was ordered. In Europe and other markets it could be ordered independently for €110 (in the UK £90). It was standard equipment for the SLK55 AMG.

In order to have the hardtop work via remote, one needed to stand within 3 m range of the car's doors, as that is where the only sensor is located and point the SmartKey directly at one of the doors handle. The remote must then be pressed until the roof has fully settled in its final position. Many people were (and still are) more than frustrated with this kind of operation and opted for an aftermarket device, called SmartTOP instead. It operates even from 100 m away by a single touch of a button and also when the car is in motion. One can even switch the car's ignition on and off during operation.

You still had to be creative if you wanted to go on an extended vacation trip

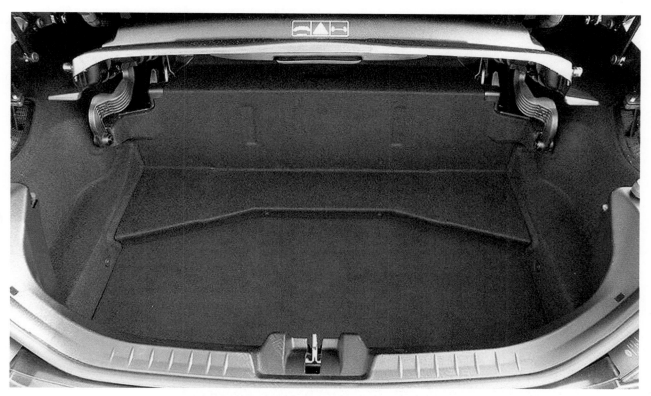

At least there was sufficient storage space with the top raised

Back in 1996, when the first SLK was introduced, the process of opening or closing the hardtop was for casual onlookers (and techies of course) fairly impressive. In 2004/2005 the sensation was more or less gone, as many more manufacturers offered such a roof for their convertibles. How did it work in real life? First of all the parking brake needed to be engaged, the solid luggage cover closed and the trunk lid closed.

Then one had to switch on the ignition and push the hardtop switch on the right side of the center console rearward and hold it in that position until the roof was completely lowered in the trunk. But before the roof started to move, the door and rear quarter windows lowered simultaneously first, with the front windows just opening a little. A few seconds later the trunk lid opened rearward.

Next, the three-part hardtop folded into the trunk, with the shelf behind the two roll bars with its two side flaps closing on top of it, forming a connection between roll bars and trunk lid. Then the trunk lid lowered itself, locking the hardtop securely in place and the front windows were raised again. All still very impressive. In the multifunction display one could then see the message: "Retractable roof open". With the R170, the opening or closing procedure took some 25 seconds, now it was over in already 22 seconds. The Ford Skyliner used in the 1950s some 60 seconds for this operation.

Naturally, the electro-hydraulic vario-hardtop remained the main USP of the SLK and according to Jörg Prigl, at that time project manager for the first SLK, it was a feature that had used up most of the engineering wizardry.

Twelve engineers had been in charge of its development and testing back in the 1990s. When regularly maintained, it is as reliable as the rest of the car. But when not cared for, the hardtop, like many other mechanical components in a car, has the nasty habit to develop over time into the car´s most troublesome feature.

The aforementioned five hydraulic cylinders, four window motors, one hydraulic pump, one relay, nine limit switches, one control switch, one solenoid valve, one control module plus tons of wiring to make it all work in sync, can make even the most hardened mechanical device fail.

5 Hydraulic Cylinders

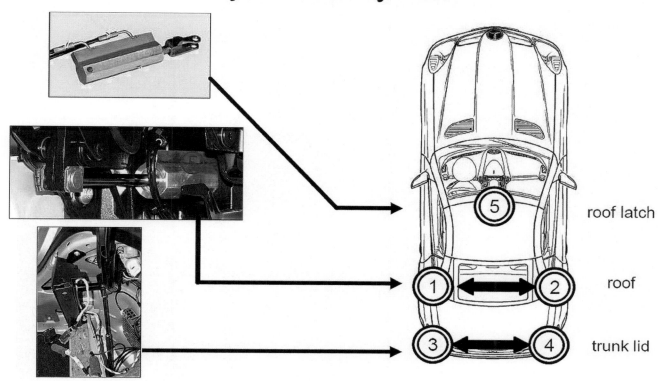

The location of the hydraulic cylinders did not change

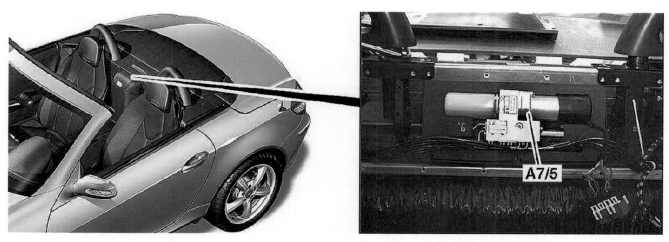

The main hydraulic unit (A7/5) is located between the seats and contains the pump motor, the hydraulic oil reservoir, one solenoid valve and the check & control valve block

Trouble shooting the vario roof

For the mechanically-challenged owner there are a few things he/she can check, before an expensive trip to the workshop might be required. This book is about the history of the SLK, and for space reasons it can unfortunately not be a guide on how to fix things mechanically or electrically. I sincerely hope that the esteemed reader will understand that for comprehensive trouble-shooting guides, there are better suited books or manuals available. The remarks and tips given here can only serve as general guideline.

If the top stops functioning, the first and easiest thing to check is the plastic luggage cover in the trunk, which should be closed in the rear position, meaning it should be hooked into the upper rear post holders. Sometimes the contact sensor for the luggage cover, which is located on the trunk's passenger side, needs a bit of cleaning to relay the message "yes, I am in position, start working". Also the car's key needs to be in the last position before it starts the engine. The SLK's manual lists four key positions from 0 to 3. That means the key needs to be in position 2. On top also the parking brake should be fully engaged and the trunk lid properly closed. One of these four issues will remove 80 percent of the reasons for the top not to function.

Without the cover firmly locked in place, the roof would not work

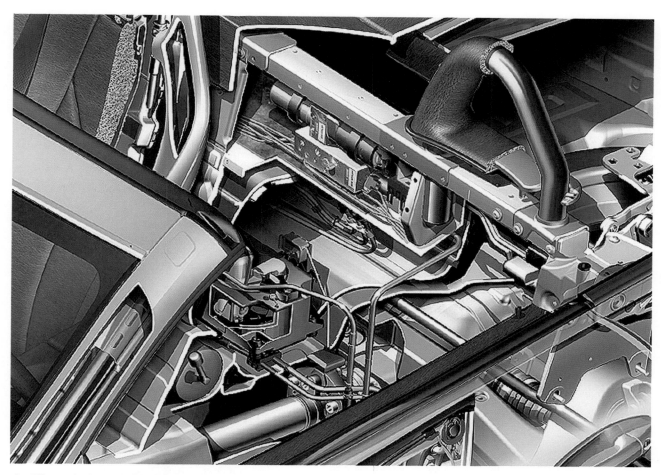

A different view at the Hydraulic pump

A fully loaded battery should have at least 11.4 volts with the engine switched off. If that is the case, the hardtop switch on the right side of the center console could be at fault. Its connections could have deteriorated over the years.

If the switch and its connections are in order, the focus should then be on the control module under a panel right behind the driver's seat (for rhd cars behind the passenger's seat). It can be seen on the next page.

As the name implies, the module controls the hydraulic unit (and tells it to change direction of pump motor rotation) and the rear side window motors. It reads switch position and sensor values and sends function and warning messages. This means, the thing is important and unfortunately quite expensive to replace if faulty (> US$2,000). Luckily it is fairly reliable, so it can help to unplug and then re-plug the module's cable.

Another focus should be the relay of the hydraulic pump, which is located in the center behind both seats. When the hardtop switch is pushed, the relay should make a clicking sound. If that does not happen, it needs to be checked with a multi-meter, whether power is being sent to it. If no power is measured, either the hardtop switch or the wiring from the switch to the relay could be damaged. If there is power, but one cannot hear any clicking sound, the best thing to do is replace the (not expensive) relay.

And even if the relay is clicking, small internal parts could have worn out over the years, so that insufficient power is sent to the hydraulic unit to operate the hardtop. Replacing the relay is also in this case a good idea and might solve the problem.

Control Module (N52) Location

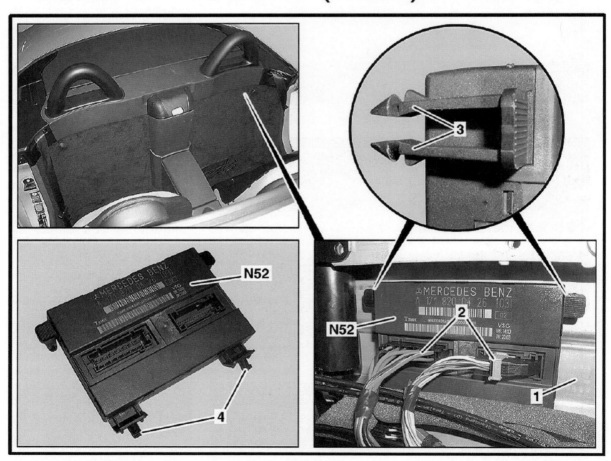

1=Cross member   2=Connectors   3&4=Mounting tabs

If the operation of the hardtop be slow, cleaning and greasing the hinges can be of help. A regular lubing of all pivot points will improve the top's operation. Also the hydraulic pump's fluid level needs to be checked. The level can be seen on the left side of the pump. The fill hole is at its very left side. In order to get to it, the hydraulic unit needs to be removed from its bracket and turned until the filler screw can be opened. But be careful and do not let oil drip through the opening. Although it rarely happens, but a low fluid level can mean that there is a leak in the system either on the pump itself or at the hydraulic cylinders. Please use only DB approved fluid such as: p/n 0009899103 or FeBi 02615.

Occasionally SLKs have issues with the top stuck halfway up. This could be caused by hardened rubber boots, which cover the sensor switches that let the tops go up and down. Replacing the boots is fairly inexpensive.

Hydraulic Unit Oil Reservoir

Checking oil level:

– Oil level must be within the marking (A) on reservoir with roof open

If level is too low, look for leakage.

Adding oil:

– Remove hydraulic unit (A7/5) from bracket
– Turn until filler screw (6) can be unscrewed without losing oil
– Add oil
– Recheck in installed position

Another possible reason for the top not to open or close completely can be a loose/hanging flap under the trunk lid. There are two of them and they are located close to the passenger compartment. Daimler-Benz offers a repair-kit for the flap problem. One is well advised to search on the internet how to use it or ask the vendor for instructions. If you are a novice, be prepared to spend the better part of three to four hours to fix it.

It can also happen that one of the door windows does not lower, when one wants to operate the roof via the remote. This can be caused by a software error in the door control unit of the remote's infrared function. Such an error can occur, when another infrared source has interfered when the roof was moving, forcing the door control unit to hang up. An easy (but unfortunately not always working) solution will be to just open and close that door that has the control unit with the software error. Meaning the side where one stood with the smart key and pointed towards the car. If that does not help, the software error can usually be fixed in the repair shop. There is no reason for an expensive control unit replacement.

And sometimes when the top refuses to co-operate, it's the good old-fashioned (gentle) shaking the top back and forth that will possibly remedy the situation, while a second person operates the button or pushes the remote button. This might help especially, when the top seals might get stuck at the windshield frame. A push on the roof near the windshield frame and proper cleaning and applying talcum powder or rubber grease (the German "Gummipflege") on the seals can be an easy remedy.

These are just some of the inexpensive ways to fix issues with the hardtop and hopefully for the wallet of the SLK owner, a possible malfunctioning can be associated with one of them. It gets progressively more expensive, when the wiring or hydraulic cylinders are at fault.

While switches, relays and cable connections can be replaced fairly inexpensively, a new hydraulic cylinder, if it cannot be repaired, is expensive. And as we already know, there are five of them. Over time their seals decay and if they are allowed to crumble apart, the particles may block the valves in the pumps. The hydraulic cylinder that operates the roof latch appears to be the one that starts to leak first.

There are numerous websites, which offer all kinds of assistance and are a great source for any information around the SLK (and of course most other Mercedes models).

I would like to mention just a few, although there are of course many others: http://www.slkworld.com is the most specific to everything SLK and that is why highly recommended. Other sites that cater to all Mercedes models and also the SLK are http://mbworld.org and http://benzworld.org in the US and www.mbclub.co.uk in the UK. In order to participate in their forums (fora), you need to register.

A look at a removed hydraulic pump

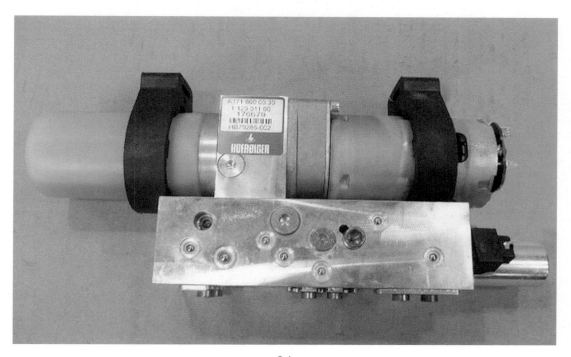

The safety features

To live up to stringent Daimler-Benz safety standards, the car came with advanced crumple zones, which Daimler-Benz had already pioneered in their basic design in the 1950s under passionate safety guru Béla Barényi, who had worked for the company from 1939 until his retirement in 1972. They incorporated full-width front and rear cross-members, which were specifically designed to help protect in offset crashes. Another cross-member near the pedals helped together with a strengthened bulkhead to keep the passengers out of harm's way. The body structure behind the seats included a convex aluminum bulkhead with another cross-member running underneath. A wall of light, yet very strong magnesium was positioned between fuel tank and trunk for additional stability and safety. The doors had reinforced high-strength steel tubes, the A-pillars had similar oval formed steel tubes welded in and the tough roll-over bars were bolted to cross members, which connected for additional passenger protection the B-pillar stubs.

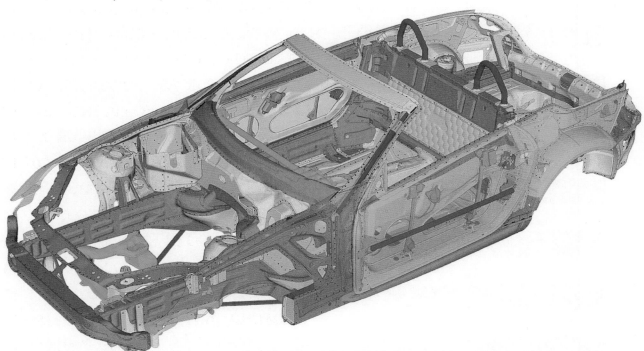

Over 40 percent of the sheet metal used for the body-shell was made of high-tensile steel alloys

The integrated restraint system included adaptive front airbags with dual-stage inflation, knee-bags and additional head-thorax airbags for increased side-impact protection. The passenger airbags would not deploy with less than 12 kg (25 lbs) in the passenger seat. Also when the optional BabySmart child seat was installed, they were automatically disarmed. Each seatbelt offered an Emergency Tensioning Device ETD, which would remove any slack in case of a collision or rollover.

Modern cars need to be light enough to save on fuel, on the other hand they need to be safe and rigid enough to offer its passengers utmost protection in case of an accident. Daimler-Benz used aluminum, magnesium and especially "intelligent" steel, which was a high strength steel-aluminum combination that was lighter than other steel materials without compromising on its stiffness or rigidity. In the R171 42 percent of all steel used is of high strength quality. It helped to ensure that the car's curb weight despite its comfort extras such as the vario-roof and extensive safety features was only 1,390 kg (3,058 lbs) in case of the SLK200. The high strength steel was also responsible for a 46 percent improved torsional stiffness with 19 percent more bending strength compared with the first generation SLK.

Another view at the High strength steel, used in the A-pillars, doors and roll-over bars

The SLK R171

The R170 had been a rather unexpected image-booster for the traditional minded Daimler-Benz company. Some even argued that the company had managed to re-invent itself with the development of the small sports car. Now after eight years and sales of 311,222 units it was time to replace the success story with a new version to keep the momentum going.

Daimler-Benz had in 1999 with the McLaren MP4/14 the most powerful and fastest Formula One race car. Management and driver errors stopped it unfortunately from scoring enough wins to hand the Mercedes McLaren team the coveted constructors' championship. That honor went to Ferrari. However, the car had a feature that was picked up by designer Steve Mattin for his second generation SLK proposal: a nose that looked very similar to the race cars and had already been used earlier by him and his colleagues on the SLR McLaren.

The 1999 Mika Hakkinen F1 car

British automotive designer Steve Mattin had joined Daimler-Benz in 1987 and had assisted also in other projects, such as the S-class, M-class and the SLR McLaren. In 1997, he received the "Designer of the Year" award by the British *Autocar Magazine* for his A-class exterior design. In 2000, he was promoted to senior design manager and is credited with the design of the ML- and GL- class. In 2004, he left Daimler-Benz after seventeen years to become head of design with Volvo.

The F1 nose must have been liked by Mattin's fellow designers too, because most of the early design proposals showed the F1 nose in variations.

When the Steve Mattin design was approved by the executive board in early 2000, designers even considered for a short time to use it for a future face-lifted SL R230. But that idea was later dropped. Anyone, who is interested in Mercedes roadsters, could get a glimpse of the second generation SLK already in January 2000, when Daimler-Benz presented at the Detroit International Auto Show the *"Vision SLA"* concept car. Based on the A-class, it was some 20 cm shorter than the SLK, but offered next to the new nose quite a few styling ideas that were further developed and later used for the R171. The second generation SLK's design process was completed at the end of 2001, so that the necessary tooling could be ordered, installed and used for the first pre-production models in 2003.

The 2000 SLA concept car

The design team and their finished project. Steve Mattin is the second from left

The R171 was finally launched on March 1st 2004 at the Geneva Palexpo, the annual Swiss international automotive show that had seen so many Mercedes car launches, among them the pagoda SL and SL R129 to name just a few. Two versions were initially available, the four-cylinder SLK200 Kompressor (internal code 171.442) and the V6 SLK350 (internal code 171.456). From Sept. 2004 onwards, also a V8 made it into the SLK: the SLK55 AMG (internal code 171.473). Gone were the days of the somewhat tame forms of the first generation SLK with its slightly feminine, cute car design. The new SLK showed on the other hand the same basic design characteristics that had made its predecessor already so popular: long hood, wide doors and a short rear.

A few more studies, before the final version was agreed upon

Like every other Mercedes, the SLK had to undergo various text cycles

If Daimler-Benz wanted to impress with one of the ugliest prototypes, they succeeded

The R171 with predecessor and ancestor

But here ended the similarity with the R170, Instead the fast contemporary, more sharply angled center axis of the hood grabbed like the nose from the SLR McLaren supercar theme. This muscular appearance was further enhanced by two horizontal wing sections in the radiator grille and the front apron, which offered a large lower air intake, vertical vanes and spoiler edges, which should help to direct air flow. In order to reduce the drag coefficient to just 0.32, also the windscreen frame was more steeply raked.

Initially the bulbous three-pointed-star front was not universally appreciated (that was the same in 1963 with the roof of the pagoda SL), but it helped to make the new car come across more aggressive than ever. The swept-back A-pillars enhanced the roadster's dynamic silhouette, while the lines of the hood continued above the widely flared wings along a slightly raising sideline and helped to create a visual link with the rear of the car, defining the wedge shape of the car's body. The lines continued over the trunk lid's outer areas, ran along striking taillights with LED technology before they ended at the rear apron, where they met with twin oval-shaped chromed exhaust tail pipes. All of this gave the SLK a much more masculine, dynamic look that differed sharply from the R170. Nevertheless, whether top up or down, the visual appeal remained as a small grand touring roadster rather than a hardcore sports car.

In order to offer more interior room, the wheelbase was extended by 30 mm from 2,400 mm to 2,430 mm (95.7 in). The body was lengthened by 72 mm to 4,089 mm (160.7 in) and widened by 65 mm to 1,777 mm (70.4 in). That gave both passengers more elbow- and knee-room with sufficient seat travel even for taller drivers. But the car was still considered small, when parked next to a SL R230.

The stiffened chassis came again from the C-class, this time the W203. It was shortened by some 285 mm (11.2 in) and used a new independent 3-link coil suspension with anti-dive geometry and stabilizer bar at the front. This new axle technology revolved around two individual link elements, which served as torque and cross struts. A third front axle link was the track rod, which connected the laterally positioned steering gear to the wheels. The spring struts consisted of coil springs, twin-tube shock absorbers and compact head bearings. An anti-roll bar was linked to the struts through a torsion bar linkage. Engine bearings, steering gear and the lower components of the front axle were all connected to an assembly carrier, which was bolted directly to the SLK's body. This did not only simplify assembly, it offered more importantly safety benefits in case of a front-end collision, as the assembly carrier would crumple and absorb part of the impact energy.

Steering and suspension were again borrowed from the C-class

43

A closer look at the left front suspension

The independent rear suspension was slightly modified for the SLK

And a more detailed look at the left rear suspension

At the rear a 5-arm multilink independent suspension with separate coil springs and shock absorbers was used. It came with a stabilizer bar, which was attached directly to the body, plus geometry for anti-squat and alignment control. An optional sport suspension was for the first time available, which lowered the ride height by 10 mm. It came with shorter, firmer and pre-compressed coil springs and sport-tuned shock absorbers. In order to have a sportier ride, the owner did not have to opt anymore for the more expensive AMG Sport Package.

Finally the somewhat mushy re-circulating ball set-up of the R170 was changed against a much more communicative rack-and-pinion system with integrated hydraulic damper. It was directly mounted to the sub-frame and offered 2.9 turns lock-to-lock. The car came standard with electronic stability control ESP, which was extremely effective at higher speeds, but almost invisible during normal driving operation. It must have been reprogrammed, because in the R170 it tended to come in a bit too early, leaving no room for the driver to "hang out the rear" in a sharp corner.

The car's stopping power came from power assisted all-disks ABS brake components, which had been taken from the E-class and delivered stops from 100 km/h in an average distance of 38.0 m (124.6 ft). The SLK350 for example came with four-piston calipers and 330 mm (13.0 in) perforated and ventilated disks at the front and twin calipers with 290 mm (11.4 in) solid disks at the rear. Brake Assist System BAS, traction control and Acceleration Skid Control ASR were standard on all SLK's.

The SLK200 came with newly designed seven-spoke 7J x 16 ET34 alloys front and rear with 205/55 R16 tires. The more powerful SLK350 offered ten-spoke 7.5J x 17 ET36 alloys with 225/45 ZR17 tires (front) and 8.5J x 17 ET30 alloys with 245/40 ZR17 tires (rear). Z-rated tires provided optimized road handling capabilities especially in fast corners, but came at the prize of slightly reduced ride comfort and increased tire noise. Daimler-Benz obviously recognized that the short-wheel car at stock ride-height could be somewhat nervous under certain driving conditions and tried to redress the balance by fitting wider tires at the rear. This issue seemed to be less apparent with the SLK200, but the larger SLK350 tires were optional.

Due to the folding top and the car`s compact size, there was no room in the trunk for a traditional spare tire. The SLK came in some markets with a collapsible tire and an electrical air pump to inflate it. Its benefit was not only its more compact size but also that it weighed some 30 percent less than a standard tire. After usage it could be deflated again, but in case of damage it could not be repaired.

Both tire and pump could be found together with the jack and wheel wrench under the trunk floor. For other markets or at customer's specification, the SLK had a TireFit repair set instead, which had been developed by Dunlop. The fuel tank was enlarged from 60 l (15.8 gal) to a more respectable 70 l (18.5 gal).

The F1 inspired front theme continued its racing heritage in the interior with two magnesium-framed bucket seats. They were shaped like race car seats, thinly upholstered with fabric or leather. Seat cushions and backrests were designed as all-foam cushions. Steel spring cores were as already in the R170 history. This became necessary as thickly padded steel spring seats could not have been accommodated in the cabin.

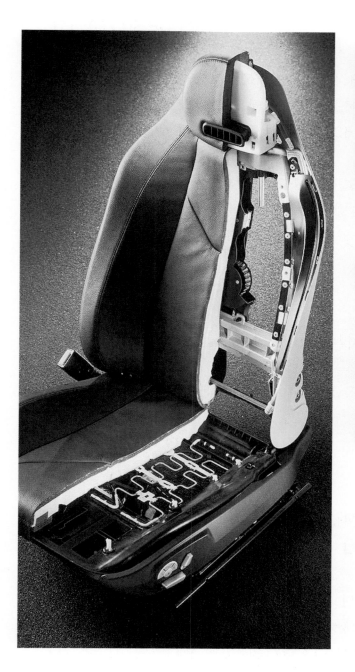

The backrest structure was made of robust magnesium

47

The center console armrest/cover was double-hinged, so it could be opened from both driver and passenger side. It could be locked with the SmartKey central locking function, which also included the glove-box cover. In most of Europe, the cars, except the SLK55, came with fabric seats. Leather was available at €1,220 for standard leather or at €1,570 for premium Nappa leather. Seat heating cost an additional €340.

The manually adjustable seats, although not limousine-size, were quite comfortable like in a grand tourer, yet also supportive enough to be used in a sports car. Eight way electrically adjustable seats were still part of the options list at €1,750, even for the SLK55 AMG. This price included electrically adjustable steering wheel and exterior rearview mirrors. But it were not only the seats that were redesigned, the whole interior looked much more modern. The instrument cluster had a driver oriented Alfa Romeo feeling to it with its two big main dials looking at the driver through silver tubes. A multi-function display was positioned between them and could be scrolled via steering-wheel controls. One of its features was a maintenance system, which showed the remaining distance to the next maintenance service, it indicated the type of service due and provided an automatic reminder about one month before that service was required.

Gone was the semi-retro look with white instrument fascia. The gauges were black with white markings. Newly designed handbrake lever and cylindrical door pulls looked firmly anchored with window switches moved from the center consoles to the upper side of the door handles. The dashboard was not anymore color coordinated with the rest of the interior. It was soft touch plastic in an all black affair with contrasting switches and push buttons with a silver-colored surface. All the grab handles, buttons or switches were well damped and delivered a quality feeling that had been absent from Mercedes cars for some time. All cars had rubber-studded brushed aluminum pedals and stainless-steel doorsill trim with Mercedes-Benz insignia (SLK55 had AMG insignia).

Also the steering wheel had been redesigned and could for the first time tilt and telescope. Tempomat with "Speedtronic" was standard on all models.

The easy to manipulate dual-zone climate control "Thermatic" was standard for the SLK350 and SLK55 AMG, but a €1,392 option for the SLK200, except for countries such as the UK or Australia, where it was standard equipment also for the four-cylinder car. Fully automatic climate control "Thermotronic" cost an additional €609 for the two larger cars or €2,015 for the SLK200 without air-con fitted.

The optional DVD-based navigation system will be covered separately in the next chapter "The COMAND system". The fully equipped cabin also boasted an AM/FM stereo with in-dash single-CD player and Tele Aid assistance for North America. The audio system's panel integrated controls and displays for radio, CD player, optional 6-CD changer and digital telephone. Fiber-optic network for high-speed communication between optional systems such as CD changer, cellular phone or Sirius Satellite Radio (North America only), came pre-installed for all cars.

The driver was greeted by an almost dizzying array of buttons and displays in the center console and on the steering wheel. Luckily they came in different shapes and were organized in logical, easy to follow clusters. So it was not that difficult to get quickly used to them

Clear control dials made the THERMATIC air-con easy to adjust

The automatic climate control "Thermotronic" cost an additional €609 for the more powerful models that had the Thermatic as standard

Daimler-Benz must have felt the gap between the SLK200 and 350 models too big, because in April 2005 a new model was introduced without much fanfare: the 3.0 L SLK280 (internal code 171.454), which shared its engine-genes with the larger SLK350. In some markets, it was called SLK300. After the 2008 facelift, it was called also in North America SLK300. From March 2009 onwards, it was named this way in all markets, including Germany. In some markets such as Australia, it was dropped in mid 2008, but was introduced again as SLK300 in April 2010. For simplicity reasons the car will be called from here onwards only SLK300. It rode on slightly smaller tires than the SLK350: 205/55 R16 tires on 7 inch five-spoke alloys at the front and 225/50 R16 tires on 8 inch alloys at the rear. Also the ventilated rotors of the front brakes were reduced from 13.0 inch to 11.8 inch. They did not come with perforations.

Initially there were 12 exterior colors to choose from. Solid colors like Mars red, black and alabaster white were available at no extra charge. On top, there were eight optional metallic colors available with Pewter (723) only available for the launch edition. Caspian blue (950) was a color that was reserved only for the SLK55. With the entry of the designo line the color options would be further increased. Those will be discussed in some more detail a bit later.

Naturally the cars did not come cheap. The entry-level SLK200 started in the UK at £28,525, while it cost in the rest of Europe €33,520. The SLK300 was in the UK available from £31,480 onwards, in other parts of Europe it started at €39,400. In the US it cost $41,600. The SLK350 cost in the UK £35,150 and in the rest of Europe €43,380. In the US it could be yours from $46,220 onwards. Add DVD navigation, Harman/Kardon audio or AMG Sport Package to this and one easily spends some $10,000 more.

The R171 continued an R170 tradition: silver was by far the most popular color

The interior was of high quality and all switches within easy reach

The engines

The second generation SLK was offered first with three different engines, a 1.8 L four-cylinder, a 3.5 L six-cylinder and a 5.5 L V8 powerhouse, which will be covered in a separate chapter. A fourth version, a 3.0 L six-cylinder, was later slipped in somewhat quietly.

Let us start with the smallest one. The SLK200 Kompressor was available in all markets except North America. Although both old and new SLK200 Kompressor looked power wise the same on paper (163 hp at 5,500 rpm), the new car had a completely revised engine (M271 E 18ML with internal code 271.944). It was a member of the new TWINPULSE generation of four-cylinder power plants, which had seen service already in the C-class since 2002. It had a bore of 82 mm and a stroke of 85 mm and was with 167 kg (367 lbs) some 18 kg (39.6 lbs) lighter than its predecessor M111.

The 1,8 L engine was made more efficient with its new "Twinpulse" *technology*

The SLK200 Kompressor engine

The engines had a cast aluminum block, aluminum DOHC cylinder heads and four valves per cylinder. They were designed to meet stricter emission regulations, reduce fuel consumption and offer at the same time an increase in output, smoothness and amount of torque provided. This was achieved by reducing the engine's friction losses and improving its thermodynamic efficiency.

The revised Eaton-style kompressor used lower engineering tolerances and was built with better rotor coatings. Driven by the crankshaft, it allowed the engine to operate with wider throttle opening, which in turn increased its efficiency. Other improvements were a cylinder head that came with variable valve timing and double adjustable overhead camshafts. The engine was also made quieter with twin contra-rotating Lanchester balancer shafts. It featured SFI fuel injection and fracture-split forges steel connecting rods.

Although the car was some 25 kg (55 ft-lbs) heavier than its predecessor, it offered a better acceleration performance (7.9 vs. 8.2 seconds) and a higher top speed (230 km/h or 143 mph vs. 223 km/h or 138.6 mph).

Its maximum torque of 240 NM (177 ft-lbs) was available at 3,000 NM (SLK200 R170: 230 NM or 169.6 ft-lbs at 2,500 rpm). Fuel consumption was an average of 8.2 l/100 km (28.7 mpg). This looked fine on paper and many were happy with the performance, but a more engaged driver would soon look for a better suited alternative.

All of this was compounded by the fact that the exhaust note was as dull as it has been already with the R170 four-cylinder. On a more positive note: the SLK200 came standard with the new six-speed manual, which was appreciated for its short shifts. Alternatively, also a five-speed automatic could be ordered. The more modern 7-speed 7G TRONIC was not available for the smallest SLK.

For those, who found the SLK200 too tame, the SLK350 was the perfect answer. The old M112 3.2 L SOHC 3-valve-per-cylinder was replaced for the SLK350 with a new six-cylinder engine (M272 E35, internal code 272.963), which increased its capacity from 3,199 cc (185.2 cu in) to 3,498 cc (213.5 cu in). It had a bore of 92.6 mm and a stroke of 86 mm. This 24-valve 90 degrees V6 with chain-driven dual overhead camshafts used as a first in a Mercedes engine continuous variable intake and exhaust valve timing. This feature did not only help to boost power to 272 hp at 6,000 rpm and to increase torque to 350 NM (258 ft-lbs), available over a wide band from 2,400 to 5,000 rpm. It also assisted in saving fuel and reduce carbon-dioxide emissions.

However, ingenuity did not stop there: magnesium two-stage resonance intake manifold increased the intake runner length at lower rpm for optimized performance and electro-pneumatic turbulence flaps at the start of the intake path increased the airflow speed at midrange rpm for a stronger and more complete combustion.

Actually, they did not just increase the airflow, they varied the path and flow of the incoming air. At lower engine speeds, the path was made longer in order to help speed-up the air intake for better cylinder charging. At higher engine speeds, the intake path was shortened in order to maximize the air volume.

The V6 with variable camshaft control for the intake and exhaust side of the engine

The engine came with one high-energy ignition coil and spark plug per cylinder. An integrated sequential multipoint fuel injection and ignition system with Bosch ME9.7 engine management offered detailed individual cylinder control, including fuel spray, spark timing and anti-knocking. Like in the M112, engineers had installed a balance shaft in the engine block between the cylinder banks in order to limit vibrations.

If this sounded all too technical, the potential owner only had to take a seat in his new prized possession, turn the key and let it all come to work. 87 percent of peak torque was already available at 1,500 rpm and one could feel that everywhere in the rev band there was nothing to complain about. Gone were also the days of that uninspiring four-cylinder SLK200 exhaust note, the SLK350 produced a wonderful throaty, edgy sound that grew even more noticeable towards the 6,400 rpm red line. Like the SLK200 the car came standard with a newly developed, responsive six-speed manual, which had lost its dual-rod system in favor of a much more direct single shaft rod-type linkage. The reverse gear had now moved from its usual left rear position to the front left alongside the first gear. With its compact throws and mated to the standard 3.27:1 rear axle ratio, it was a great deal of fun to push the car from zero to 100 km/h in a mere 5.6 seconds or from 60 to 120 km/h in 8.7 seconds. The quarter mile was consumed in 14.1 seconds at 165.7 km/h (102.9 mph) and the top speed was electronically governed at 250 km/h (155 mph). That was almost former SLK32 AMG territory.

For the more comfort oriented driver, the car could also be equipped with an industry-first seven-speed automatic transmission 7G-TRONIC, which was fitted standard to the SLK55 AMG and SLK models with AMG Sport Package. For some markets such as South Africa, it was the only transmission offered. It came optional with gear shift buttons behind the steering wheel and its touch-shift feature allowed for manual shifts by nudging the gear lever either left or right from the Drive position. One of its hallmarks was its excellent use of its gears.

Unlike other transmissions, which run sequentially through their gears, while up- or downshifting, this new transmission can skip as many as four gears at a time, which resulted in much faster up/downshifts, coupled with the ability to get back on the power much sooner.

The man behind the new engine technology: Anton Waltner

For North America, the manual shift was only available in the 2005 model, the car's first year in that market. Due to low demand (although it was favored by many automotive journalists) it was dropped the following year and the 7G-Tronic became standard equipment.

The SLK300 used a downsized 3.0 L version of the 3.5 L engine (M272 E30, internal code 272.942) with a bore of 88 mm and a stroke of 82 mm.

It offered 231 hp at 6,000 rpm with 300 NM (221 ft-lbs) of torque, which were available from 2,500 to 5,000 rpm. It accelerated from zero to 100 km/h in 6.3 seconds and had its top speed again electronically limited at 250 km/h. With a curb weight of 1,440 kg (3,160 lbs), it was some 25 kg lighter than its bigger brother. The engine was different from the one used in the W203/204 C-class. Contrary to them it could not use E85 (85 percent ethanol) fuel.

M272 balance shaft and other issues

Early SLK280/300/350 versions had occasionally issues with the balance shaft sprocket (M272 engines produced between April 2004 and September 2006) as the drive gear was not in all cases sufficiently hardened. I use the word "occasionally" on purpose, because many owners did not experience any such issues and some reviewers even scolded me for talking about it at all. Yet, it has happened, it was widely published in the press and that is why it should be discussed here.

In a 90 degrees V6 engine, engine vibrations are inherent and the balance shaft is used to reduce those vibrations. It is basically a shaft inside the engine block, which is positioned parallel to the crankshaft with offsetting weights. It is driven by the center idler sprocket on the timing chain. This sprocket could wear out prematurely. If found faulty, it cannot be replaced in itself, as it is just a forged part of the balancer shaft. That is why the whole balance shaft needs to be replaced. This problem does not happen with the four-cylinder and eight-cylinder SLK cars.

Here is a related article that has been copied from Wikipedia (all copyrights are with Wikipedia). With regards to the engine number, the last eight digits are important:

M272 engines that were sold between 2004 and 2008 with engine serial numbers below 2729.30 468993 often show early wear of the balance shaft gears, requiring extensive repairs at a retail cost of over $4000. These com-plaints led to a class action lawsuit against DB (Greg Suddreth and Paul Dunton v. DB USA, LLC), which alleged the M272 engines are equipped with defective balance shafts gears which "wear out prematurely, excessively and without warning, purportedly causing the vehicles to malfunction, the check engine light to illuminate and the vehicle to misfire and/or stop driving." The suit further alleged that DB knew of this, sending out repair bulletins on how to address this issue and ultimately changing the balance shaft gears to avoid this problem. This suit was dismissed with the judge agreeing with DB that because the gears fail at 60 -80K miles and outside of the warranty period, DB is not legally responsible for these problems. A second class action lawsuit is being organized.

Having now early 2017, that lawsuit is of course no more. As the judge at that time said, if it is outside of warranty, it is usually the owner's problem. The case was also so poorly prepared by the filing lawyers, with so many facts apparently wrong, that it was justified to have it dismissed.

All this does not help the SLK owner, who is concerned, whether his car might develop similar problems, although now in 2017 most of these issues should have been remedied. Nevertheless, let us start with a worst case scenario: if you have a model 2005 to 2007 car and the „check engine" light on your instrument cluster illuminates, have the engine checked by your trusted repair shop immediately with an ECU (Engine Control Unit) code reader.

If the following ECU code scenario is found: *FC (Fault Code) 1200 and FC1208 (OBDII or "On-Board Diagnostic" codes P0017 and P0016),* do yourself a big favor and DO NOT WAIT!! Your trusted mechanic will tell you that your engine's balance shaft needs replacement asap.

In most cases code P0017 will show up first, followed intermittently by P0016. Eventually both will show up permanently. A quick search on google for ODBII or OBD2 codes will give you a lot of information, what these codes are about.

A detailed look at the location of the balance shaft in the center of the photo

Worn out vs. new

The fault codes 1200 and 1208 are stored in the ME-SFI (Motor Electronics Sequential Fuel Injection) control unit's fault memory. Even when the fault memory is erased and the engine is started again, both fault codes will reoccur immediately. This is due to the positioning of the camshafts of the right cylinder bank relative to the crankshaft being impaired by the worn sprocket.

All this sounds terribly technical unfortunately, but balance shaft sprocket failures are not that common and can be fixed of course. Balance shaft kits cost around US$430 or €300. But for this the engine has to be taken out and other parts might/will have to be changed too in the process. To take out the engine, dismantle it, repair it, put everything back together and re-install it in the car will take at least some 21 man-hours. So expect the total job to be well in excess of US$4,000. It is still cheaper than wait and have the entire engine later replaced. If you own a model 2008 or later car, those issues have been remedied. If you are not sure, have your VIN checked for your engine number. This can be done either through your repair shop or one of the most helpful Mercedes dedicated websites that have been mentioned already earlier. In case you have no idea what your VIN is all about, it has a separate chapter later in this book.

As already mentioned earlier, it only affected engines up to engine number 30-468993 (counting the last eight digits). So if you have an engine that falls with the last six digits below said number, but starts with 31 on its last eight digits instead of 30, you are in luck. If you are the second owner of your SLK, chances are you are not even aware that such a problem had or could occur. You also might not know whether any of this has been fixed already. The best way to find out is to go to your repair shop, which can read Daimler-Benz ECU codes. Have them pull out the codes to make sure, you do not have a 1200/1208 DTC (Diagnostic Trouble Code) stored, even if it has not yet illuminated the MIL (Malfunctioning Indicator Lamp), which is another expression used by mechanics for "check engine light".

In most cases, the ECU reader will not see these codes, because it has never happened with your particular car and there are good chances that it will never happen. A balance shaft is like piston rings or valves a moving metal object that will show natural wear over time. From talking to mechanics and reading many related articles on the net and in forum groups, "our" balance shaft sprocket issue is a premature failure of a part that should last for more than 200,000 miles. Most such premature failures have happened between 20k and 100k.

However, if you have a car within the engine numbers range that has more than let us say 120,000 on the clock, there should be nothing to worry about anymore. If the sprocket would have been faulty, it would have caused problems already.

Why did the problem occur at all? It was not an issue of the wrong oil, wrong oil change intervals or wrong assembly, it was finally traced in 2008 by Daimler-Benz to a wrong heating/cooling treatment during the manufacturing process of the sprocket, which was made from sintered steel. It appeared to be an error that occurred inter-mittent and did only affect a few batches. And that was most likely also the reason, why it could occur over a fairly long period of time. In the meantime, the sprockets are made from conventional steel.

Unfortunately, Daimler-Benz does have no way of knowing exactly, how many of the 550,000+ engines were affected (the engines were used across the entire Mercedes model line), but they assume according to a German magazine (*Auto Bild*, August 2011) that less than one percent of all engines were affected. They only know the last engine that was possibly built with incorrectly forged parts in September 2006. It was #30-468993.

As I already said earlier, we can assume that the issue has been addressed and fixed by now on all cars that were affected. Chances to have the balance shaft repaired in 2017 by a Daimler-Benz dealer on warranty are dim to say the least. They will just tell you that the car is already too old for any warranty claims. On top, dealership sales staff can be pretty ignorant of issues like this and if they are aware, they will most probably not let you know.

On the other hand, one cannot really blame them, as many of them, with exceptions of course, are sales people, not "gear heads" like us. They are being trained just enough to sell cars and when we enter a showroom and talk to one of them, it can happen that we know more about the car in question than they do.

Realizing all this, it might be a good idea, if one is unfortunate to have an affected car, to contact Daimler-Benz headquarters in your country directly WITHOUT going through your local dealer. This is especially important, if your car has not been serviced regularly by your dealer and he does not know you that well. Talk to customer service directly in a nice and polite way or send them a nice email. Asking them does not cost anything. I have read somewhere that the highest chances for help are at the beginning of a fiscal year.

By the way, the balance shaft sprocket problem occurred also with the M273 5.0 l V8 engine, but that was not the one that propelled the SLK55, which used the M113 engine. Other issues with the M272 engine were camshaft solenoid problems (which are relatively easy and cheap to fix) and worn piston rings (mostly on 2005 models), which caused excessive oil consumption.

Again, most of these issues must have been repaired by now. If you plan to buy an SLK with M272 engine, ask the current owner to show you the Master Vehicle Inquiry. That is the data sheet from which one can see, whether the car in question had all the work done by Daimler-Benz or an authorized repair facility. It also shows, whether the car has been in for oil problems.

The COMAND system

In order to keep the passengers entertained and informed, various optional audio equipments could be ordered. The three non-Harmon/Kardan units came with a total of nine speakers, three in each door, a center-fill speaker in the top section of the dashboard and two additional speakers behind the passengers. The first such system was called Audio 20 CD stereo radio and came with telephone preparation at €750. It had no navigation function. Some markets had this system standard for the

SLK 200 and SLK280. In North America it was standard for all cars, including the SLK55 AMG. It featured an AM/FM/LW audio system, which allowed ten station presets per waveband. It came with an integral single-disc CD player and four-channel amplifier, which produced an output of 4x25 Watts. For FM stations it offered Radio Data System RDS and a station scan function. An optional six-disk CD changer was available for €460 and installed in the glove box compartment.

Contrary to the more expensive systems, the Audio 20 had its CD slot below the screen

More expensive at €2,140 was the Audio 50 APS, which was not available for North America. It included a twin-tuner FM receiver, CD drive and an arrow-based navigation system with dynamic route guidance, based on RDS-TMC. The information appeared on a 4.9-inch color display with graphics capability. For Europe for example the route network of the main countries was stored on a single navigation CD. The CD drive was capable of playing music from a CD, while the navigation function was activated.

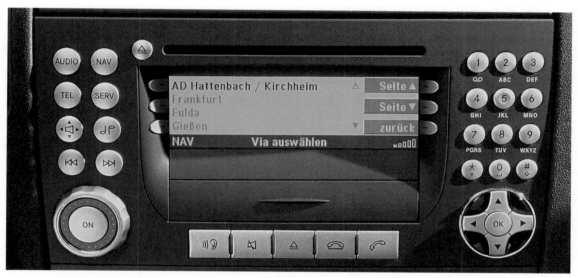

Audio 50 APS car radio was fitted with a navigation system and dynamic route guidance.

COMAND APS (here NTG 1) was identifiable by its 16:9 format color screen

On top of this a new COMAND APS could be ordered for €3,240. It came with a CD player (MP3 capable) and a separate DVD drive for the navigation system.

In case the soon-to-be SLK owner wanted a better audio system, he/she could opt for one of two entertainment packages. Both came with a 380-Watt multi-channel Harman/Kardon LOGIC7 digital surround system with 11 speakers. A microphone inside the SLK detected the actual noise level and enabled the system's micro-computer to adjust volume and sound level.

This worked even, when the top was down and offered both passengers perfect surround sound through two additional speakers, which were located above the stowage compartment on the car's rear wall. The first entertainment package came without, the second with the navigation system. For North America it was part of the Premium 2 package w/o navigation or Premium 3 package with the DVD-based navigation (both are covered in the next chapter). In some other markets it was for example for the SLK350 part of the designo package,

The optional CD changer was located in the glove box compartment

The name "COMAND" covers all infotainment systems of Daimler-Benz since 1993 and had been used first in the new W140 S-class. Its first name was "Communications-and Navigations-System" and was the first fully integrated telematic-system in the automotive industry.

In 1998, the name COMAND had been adopted. It stands for "Cockpit Management and Data System" and has been upgraded continuously over the years. The first such system was the COMAND 2.5 and was produced together with its successor COMAND 2.0 and COMAND 2.0 MOPF by Blaupunkt. Collectively all three systems carry the name COMAND 2. When production was changed to Siemens/VDO, the previously used D2B audio network was replaced with a much faster system, called "Media Oriented Systems Transport" MOST. It was now called COMAND APS NTG1 (New Telematic Generation). Part of the navigation was the "Auto Pilot System" APS, which used for route calculation and latest traffic information the "Traffic Message Channel" TMC. This COMAND APS NTG1 was first seen in autumn of 2003 in the W220 S-class and the face-lifted C215 CL-class for model year 2004 and from 2004 in the R230 SL for model year 2005. Until 2009 it was also fitted in the W211 E-class and the W219 CLS and of course the R171 SLK.

It was produced for some Mercedes models by Becker for European markets and was the first DVD based system to carry the whole of Europe on one DVD. It could also carry all of North America, including Hawaii. It could display so-called points of interests, such as fuel stations, restaurants, hotels etc. For the R171 it was produced for all markets by Siemens. It offered a 6.5-inch color display in 16:9 format and came standard in some markets for the SLK350 and SLK55 AMG.

An auxiliary audio input (at the right side of the glove box) was available for external devices such as an iPod, when the Daimler-Benz iPod interface was installed. Analogue TV reception was optionally offered, but would only work if the car was stationary. From 2005 onwards also digital TV reception (DVB-T) became available. Various after-market suppliers offered software that let the TV also play while on the way.

The NTG1 system consisted of three separate devices, which were previously all housed in one unit. The head unit COMAND was of course with the display in the center console, it could play CD discs with MP3 music.

The navigation processor was installed in the SLK's trunk (together with the multiple handset interface), while the audio gateway AGW was located in the passenger foot area. It contained the FM/AM tuner, amplifier, MOST bus master controller and gateway for diagnosis and configuration.

The Harman/Kardon premium audio system, CD changer, TV tuner and telephone system with voice recognition all connected to the MOST bus in order to route the system's audio in/out. With the Harman/Kardon system installed, its amplifier replaced the default AGW with its built-in audio gateway. COMAND units installed in cars destined for North America, Japan and China, would not work in Europe and other parts of the world (and vice-versa), as the maps could apparently not be loaded, even when the old ones had been deleted.

The COMAND APS NTG1 used proprietary map disks, which could be bought only at Daimler-Benz dealers or otherwise approved outlets. They were made by either Navteq or Tele Atlas. Daimler-Benz used a colored flash on their map disk packaging, NTG1 disks had a green stripe.

For the SLK R171, Daimler-Benz changed in April 2008 with the car's facelift from the NTG1 to the similar looking NTG2.5.

This was a cost-reduced system, as it put all components including the navigation processor into one double-Din unit and thus simplified the wiring. The map disks used now a blue stripe.

The new unit came with inbuilt blue-tooth phone systems with the map data now stored on an internal hard drive. It also included as a first for the SLK "Linguatronic", which was a voice-operated recognition system. It could also be ordered with the Audio 50 system and worked in most conditions without flaws. You told the system your destination and it would save you valuable time trying to figure out how to enter the address manually. It would also recognize names from your phonebook or radio stations.

The NTG2.5 system came with a DVD changer that could also play MP3 CDs or DVDs. It supported the Media Interface and worked with the optional iPod kit ($375), where control and track display could be controlled from the steering wheel and the instrument cluster. The inbuilt memory card slot could not support cards with over 2GB storage capacity.

The "all-in-one" NTG 2.5 was used after the facelift and offered a memory card slot on top of the CD/DVD slot

In order to update the firmware, which Daimler-Benz released as so-called telematic disks, you were required to visit the Mercedes dealer or any authorized outlet with Star Diagnosis facilities. To find out, what firmware version is installed in the COMAND unit, this is what you have to do: Please enter first the unit's "engineering mode". To accomplish this, press NAV, then press (gently) the following thee buttons for some 10 seconds: 6, *, and Phone-Hang up. Now press OK, repress OK one more time and then the required information will be shown on the display. It should read something like 02/43/09/29/00 or similar. The 09/29 part is the version of the firmware used in the unit.

To fit the NTG2.5 into a R171 that came with the Audio 50 system or without navigation is tricky, but it seems that it can be done. A search on google or with one of the earlier mentioned Mercedes enthusiasts websites will give some more input. If the car came equipped with the Harman/Kardon audio, you would need a new amplifier and special adapter looms. There are two caveats though; the first one: if the unit came from another car, it would display: *"Antitheft protection activated, please visit your Mercedes-Benz workshop"*.

This would happen as the unit would read the chassis number from the ignition switch, once the ignition has been turned on. Naturally the previous car had a different chassis number, so the PIN code that came with the NTG2.5 system, was required, which had to be entered using Start Diagnosis. And the second caveat: I have read on some Mercedes forum discussions that some Daimler-Benz dealers have apparently problems with updating the firmware of later installed units, as the Star diagnosis does not recognize them.

With the COMAND upgrade, Harman/Kardon surround systems came with Dolby 5.1/DTS and an increased output of now 500 Watt (previously 380 Watt).

Another interesting add-on (at least for North America) was the optional Sirius satellite radio (part of the Premium 1 package), which integrated with the standard audio system and controls. It came as a subscription and offered over 100 all-digital coast-to-coast streams, 60 of them for commercial-free music of all kind.

The option packages

Daimler-Benz had various options for different countries and to include them all in this book would make reading somewhat tiresome. Most packages that were offered for cars sold in North America did not make it to other countries, as they were marketed there not as packages but in many cases as individual options, which could be loosely combined or as an example in case of the AirScarf, ordered only, when the car was equipped with heated leather seats. This chapter will concentrate first on the packages that were available for the North American market, where a total of nine different packages could be ordered for the SLK (four for the SLK55). At the end of this chapter, the main European options will be discussed.

The first package mentioned, the **AMG Sport Package**, was available for all markets and for all SLK, except of course the SLK55 AMG. It was already known from the R170 and came with special 17 in AMG five-spoke alloys, but used the SLK350 tire sizes. The lower body parts were adapted with a special front air dam with mesh intakes, special side skirts and an AMG rear apron. The lowered sport suspension made the ride a bit less comfortable, but cornering more secure, less sedan-like. In case the car was fitted with an automatic transmission, the package came with shift buttons behind the steering wheel. Other goodies included cross-drilled disks, four-piston calipers (both of these features were only included after the 2008 facelift). The AMG package did not offer the SLK55 exclusive quad tip exhaust. But as long as the car was in production, Daimler-Benz offered an AMG muffler kit with twin tips as a dealer-option. These kits (not the original AMG exhaust) seem to be hard to come by today. The package did not include interior upgrades (except the steering wheel shift buttons) and cost $4,100 for the SLK280 and $3,810 for the SLK350.

To add a bit of confusion, Daimler-Benz offered for the SLK next to the AMG Sport Package also a Sports Package and a Performance Package. All three were different options with the first two only available for the "basic" SLKs and the last one only available for the SLK55 AMG. All three will be discussed here in more detail.

SLK with (above) and without (below) AMG Sport Package

To make the cabin more enjoyable when driving top down in cooler weather, a **Heating Package** was available. It came with heated seats (standard in the SLK55), wind deflector behind the seats, plus a first in the automotive industry: the Mercedes AirScarf system, which was a fancy name for saying that it blew warm air around one's neck and shoulders from a vent in the headrest of each seat. It offered three-stage temperature control and even adjusted the flow of air depending on the car's speed. In Europe, where the SLK (except the SLK55) came standard with fabric seats, the AirScarf was only offered when heated leather seats had been chosen. A heating package, like for North America, was not available in Europe, while a separate AirScarf was for North America not available for the SLK350. It could be ordered individually for the SLK55, as heated seats came already standard as noted earlier. It cost $980.-

The **Comfort Package** became in 2007 part of the **Premium 1 Package**, covered a bit later. It offered eight-way power seats with three-position memory for each seat and a four-way adjustable steering column. The driver's seat memory included steering wheel and external rearview mirror positions. It did not control the internal rearview mirror though. The package was standard in the SLK55.

The **Trim Package**, later called **Appearance Package**, was not available for the SLK55. It offered polished Vavona wood trim on dash, doors, steering wheel and handbrake. It also came with 17 in six-twin-spoke alloys, black headlamps inlays, black roof-liner and Star and Laurel wheel-center covers. It cost $1,000.

SLK350 with elegant looking Vavona wood trim

The **Lighting Package** came with bi-xenon low- and high-beam headlamps, heated headlamp washers and corner-illuminating fog lights. These corner lights (standard on many US-built cars since the 1960s already) should help in turns or dark driveways at speeds of up to 40 km/h (25 mph). The system monitored the angle of the steering wheel and the turn-signal use and illuminated either fog light. This function was not part of the package for the SLK55 and a car with AMG sport package. It cost $1,010.

The **Premium Package** was available in three versions. The first one, **Premium 1**, came with a multitude of comfort- and driver-oriented extras, such as:

- Speed sensitive power steering
- 8-way power seats with power-adj. steering column
- Sirius satellite radio
- Auto-dimming rearview mirrors (inside and driver side)
- Remote window and hardtop operation
- Ambient interior lighting
- Entrance lights in both doors
- Illuminated center console storage
- Illuminated vanity mirror
- Auxiliary outlet in glove box
- Rain-sensing intermittent wipers
- Three-button garage door control
- Automatic climate control with sun and humidity sensors

The Premium 1 Package cost $2,250. The items of this Premium 1 package were standard in the SLK55.

The **Premium 2** package included on top of package 1 at a cost of $3,995:
- COMAND cockpit management with 6.5 in color-LCD screen
- Harman/Kardon LOGIC7 digital surround-sound system with 11 speakers
- Hands-free cellular phone

The **Premium 3** package included all of package 2 plus the DVD-based navigation system. Both packages 2 and 3 did not come standard in the SLK55. The Premium 3 package cost $5,195.

In the **Run-Flat Tire Package** Daimler-Benz offered extended-mobility tires with self-supporting sidewalls, which should help maintain the tire's structure. That way the SLK could continue driving for a certain amount of time after a tire puncture. In conjunction with this feature, a low-pressure warning function was incorporated in the multifunction display of the instrument cluster. This package was not available for the SLK55. It cost $250.

The two *Entertainment Packages* (later part of the Premium 1 and 2 packages) are covered in the chapter that deals with the COMAND system.

For **Europe and most other non-North American export markets**, many of these options could be ordered separately. An alternative to the AMG Sport Package was since September 2006 the *Sports Package*. It cost €1,650 (SLK200), €1,470 (SLK300) or €1,180 (SLK350) and came without the AMG body styling. Otherwise, it offered most of the technical upgrades such as a revised brake system with perforated brakes at the front and calipers with Mercedes-Benz letterings. New twin six-spoke 18 in alloys were added, 7.5 x 18 ET37 with 225/40 R18 at the front and 8.5 x 18 ET30 with 245/35 R18 tires at the rear. The sport suspension was lowered by 10 mm and a sport air filter for a throatier sound was available for the SLK300 and SLK350.

Other improvements were the headlamps with darker surrounds and an AMG-style spoiler lip on the trunk lid. More visible changes were introduced for the interior which came with red craftsman-like topstitching on the door armrests, steering wheel, gaiter of the leather gearshift-lever and also with floor mats edged in red. Red was also the color of needles in the instrument cluster and the seat belts. In case one thought this would be too much red in the cabin, one could order black seat belts instead (code U17). If the car came equipped with optional designo interior, the red seams were replaced with designo specific colors and the seat belts came in black. Carbon fiber replaced with greetings from the pre-facelift R170 the wood or aluminum trim, It came also above the glove compartment. The black roof liner and black trim to the left and right side of the steering wheel should serve a bit as contrast.

In case one did not want the AMG design nor the Sports Package, but just the lower suspension without the improved brakes, that was possible in conjunction with special alloys (code 639 or 782) and could be ordered for €245. One the other hand, if one liked the AMG body styling, but did not fancy the slightly harder and lower ride of the sport suspension, one could order for €3,550 the AMG front and rear apron with the side body kits. Daimler-Benz really tried to cater to almost all tastes.

As already mentioned earlier, the SLK55 AMG could be ordered with a Performance Package. That was from August 2006 onwards with the opening of the new AMG performance center. This package is covered in the chapter that deals with the first SLK upgrades. If one wanted for his SLK55 AMG additional tuning options prior to this date, one had to look actually at three different packages. The most important one was of course the **Driver's Package** that was already mentioned earlier. For €2,550, the speed limit would be raised to 280 km/h (174 mph).

The second part was the *Performance Suspension Package*, which cost €1,130 plus another €490 for installation and offered a lowered, stiffer suspension with special springs and shock absorbers.

The third package was the *High-Performance Braking System.* It cost €1,940 plus another €210 for installation. It came with internally ventilated and perforated composite front disks with six-piston fixed calipers. They were enlarged from 340 x 32 mm to 360 x 32 mm. At the rear internally ventilated and perforated brakes offered 330 x 26 mm disks with four-piston fixed calipers.

The infrared hardtop remote, which was part of the Premium 1 package for North America, could be ordered separately for some £90 in the UK or €110 in other parts of Europe. It came built into the SmartKey and worked from up to three meters away from the car. The infrared signal was picked up by a small sensor at the back of each door handle.

While the after-market remote for the R170 worked with just a push of a button, the Daimler-Benz R171 remote only worked for safety reasons, when one pressed the button continuously for around 22 seconds until the roof had fully settled in its final position. After-markets remotes such as "SmartTops", which operated by radio link, were also available for the R171, which meant one could stand further away from the car. They worked with one push of the button instead of continuously pressing it.

Like the R170, the successor could be almost tailor made from a large variety of interior and exterior colors. Three designo packages (Espresso, Graphite and Chablis) were available at €3,020 with solid or two-tone interior Nappa leather variations. For the SLK55 four two-tone interior options were available at €1,450.

Some of the interior options were only available in certain packages. If one wanted to add for example *Designo Interior* to the SLK200, one had to buy also the interior light package, which was standard in the more expensive cars. Exterior colors came in five different variations. The three already mentioned solid colors were available as a no cost option for the SLK, with metallic paints costing an additional €630. These twelve colors could be combined with six interior colors. While the sometimes intense color schemes, offered on the first generation SLK, were fashionable in the 1990s, there was now a new trend towards more subtle color options. That meant that bright color options such as yellow, light green etc were gone.

That of course did not stop designers to come up with some pretty evocative names such as "tanzanite blue" or "andradite green" and even "benitoite blue" for a designo paint scheme. Designo metallic paints (outside the previously mentioned three packages) cost €1,590 while designo varicolor/mystic colors (codes 0034 and 0036 for example) were a €2,420 option. This was topped by designo chromaflair paints (code 0031 for example), which cost a whopping €5,1450 Altogether, the designo range of appointments included twelve special metallic paints, ten leather trim shades, single tone leather or Alcantara trim and Alcantara roof lining. Special designo wood trim came in either Japanese ash or natural Poplar.

Chromaflair I: a dark-blue purple metallic (code 031)

The SLK55 AMG

There are plenty of discussions on the internet about what constitutes an AMG model. Daimler AG sold and sells AMG Sport Packages or body kits for almost any of their models. This makes it difficult for many to distinguish between a real AMG and an AMG look-alike. Also on the second-hand market, cars with the Sport Package are sometimes advertized as Mercedes AMG cars. But true AMG cars are distinguished by what is under the hood. All AMG power plants (except R129 versions) have a silver engine cover with the name of the technician, who built that very engine,

proudly displayed on top of it with a plaque. Having said that, it must be noted though that AMG started with the "name affixing" practice only in the third quarter of 2001 with the introduction of the SL55 AMG R230. In case of the SLK, it was started only in 2003, so earlier SLK32 came without them.

AMG was founded by Hans Werner **A**ufrecht and Erhard **M**elcher in June 1967 as a racing shop. The third character "G" in the AMG name did not come from the company's first location, but from Aufrecht's place of birth "Großaspach".

H. W. Aufrecht (to the left) and his team in front of the Red Sow

Erhard Melcher in the very early AMG days

Both gentlemen were former Daimler-Benz technicians, who offered initially engine performance packages and various unofficial upgrades and accessory packages for Daimler-Benz cars. Daimler-Benz engineers used to laugh at them due to their tiny business size.

They also did not take them too seriously because of their continuous underfunding problems. But over the years AMG managed to mature into a full-fledged racing company with its first major rally success in 1971, when they managed to score second overall place behind a Ford Capri at the 24 hours race at Spa with their impressive 300SEL 6.3, called "Red Sow" (the 6.3 was actually a 6.8). In 1990, AMG signed with Daimler-Benz a cooperation agreement and in January 1999, Daimler-Benz acquired 51 percent of the company. At that time, the name was changed into Mercedes-AMG GmbH (Ltd.). Since January 2005, AMG is a wholly owned subsidiary of Daimler AG. The first official model developed by AMG in 1991 was the AMG 500SL 6.0 V8 (M 119) R129 with Bosch KE injection, 374 hp at 5.500 rpm and a

maximum torque of 550 Nm (405.7 ft-lbs) at 4.000 rpm.

The question whether an AMG version of the SLK was really needed with the SLK350 around, is rather academic. It was offered already in 14 other Mercedes models, so it was only natural to continue that practice in the little sports car too. "We offer race technology in street cars for the top end of the market" was the statement of Hubertus Troska, AMG CEO at the time of the new car's presentation.

To demonstrate the car's ability, it was used as Safety Car in the 2004/2005 Formula One season, before it was to hit with an identical engine (but tamer exhaust system) the showrooms around the world. To come back to the question, whether such a car was really needed, head of AMG sales & marketing Mario Spitzner had this to say: "Naturally nobody does really need it, but on the other hand there is also no harm in owning it and to have plenty of fun with it. Our clients usually have more than one car. Many even own on top of the AMG car a Porsche or Ferrari".

Stocky and wide, or awesome with brilliant performance. The AMG SLK did not only make an impression through its design. It was a surprise that a V8 was offered in the SLK at all. No competitor had optimized the concept of a V8 engine in a smallish package with such consequence. And one could not blame AMG (or Daimler-Chrysler) to have cleverly used the opportunity to plant a superb engine in not just one, but a row of its models. And as an engineer, who would not have been

tempted to try to get this thing into the somewhat small engine bay of a tight roadster.

Sitting in a car with a curb weight of 1,540 kg (3,390 lbs), propelled by a V8 with 360 hp meant on a German autobahn that there was nothing to worry about in case someone grew overly large in your rearview mirror. Just floor the pedal and that nuisance would be gone in most incidences.

The M113 V8 was closely related to the M112 V6 and was offered in Mercedes cars with and without kompressor. It had seen service first in the C55 and E55 in 1998 and came with aluminum SOHC cylinder heads with three valves and aluminum engine blocks. Each cylinder had two spark plugs and was lined with aluminum/silicone. Other technical goodies included a magnesium intake manifold, SFI fuel injection, one piece cast camshaft and fracture-split forged steel connecting rods.

The M113 that was used in the SLK was the 5.4 L naturally aspirated E55 version (329.5 cu in). It came with light cast aluminum pistons, oil-spray nozzles to keep the piston crowns cool, variable intake manifold, variably adjusted camshafts and a newly developed twin-pipe intake system. Other engine versions were E43, E50 and ML55. The "E" stood for the German word **E**inspritzung (injection). "ML" meant that the engine was used with a kompressor, in German: **M**echanischer **L**ader.

The M113 E55 was used from 1999 onwards already in the R129 SL55 AMG

The ML55 was even the basis for the M155 in the mighty SLR McLaren. The E55 had a bore of 97 mm and a stroke of 92 mm. Its power output was available at 5,500 rpm, while its maximum torque of 530 NM (391 ft-lbs) was delivered in a wide band from 2,800 to 5,400 rpm. The SLK55 raced from zero to 100 km/h in 4.9 seconds, reached 200 km/h (124 mph) in 17.5 seconds and had like all other AMG cars its top speed of 250 km/h electronically limited.

And this limitation was for most SLK AMG owners at least in Germany the reason, why they visited the AMG headquarters in Affalterbach a second time. They asked to have the limiter removed. After some €2,550 had changed hands for the so-called **AMG Driver's Package**, it was of course granted together with the pledge to please use only approved tires and have the tire pressure religiously checked. The package included an invitation to attend an AMG "Power & Passion" driving event. Now the SLK could run 280 km/h (174 mph) and the SLK driver did not have to fear anymore that a Porsche Carrera could overtake him. As Spitzner noted: "That is quite important for some of our clients, for some even the major reason to buy an AMG tuned Mercedes at all".

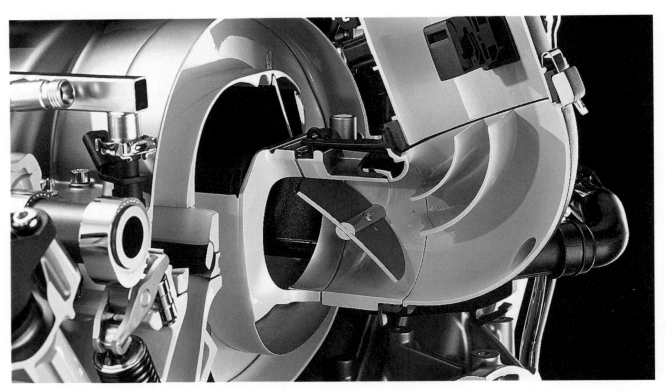

The intake module was fitted with flaps that changed the length of the intake pipes according to the engine load

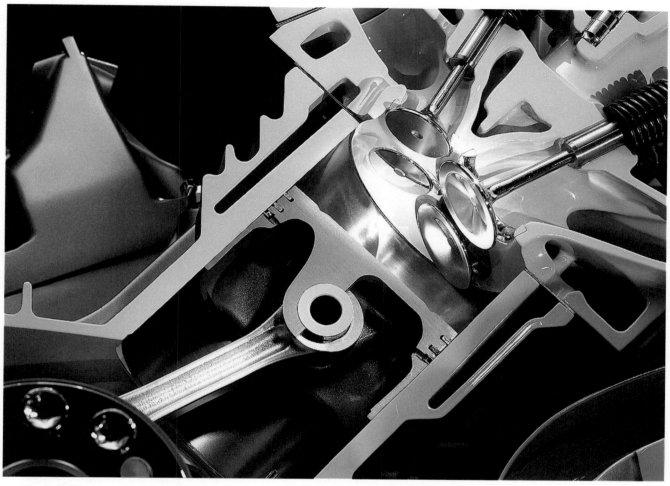

Tumble flaps in the intake ducts improved the flow in the part-load range

Naturally, the car came with various refinements like a new front apron with round clear-glass fog lamps with chrome rings. Integrated into the front apron was an engine oil cooler, separate side outlets in the front fenders allowed hot air to disappear. The radiator came with black-painted cross fins; other goodies included bi-xenon headlamps and darkened taillights.

Also new expressive side skirts were added, which made the car look more muscular and dynamic. An efficient rear spoiler lip was said to reduce lift over the rear axle by almost 40 percent. A specially developed sports exhaust system with two sets of twin tailpipes and AMG exhaust tips poked almost menacingly from the rear.

Finally an SLK AMG that is not in silver or black

The suspension was made stiffer with specially adjusted damper struts and anti-roll bars. For better fast cornering abilities the chassis was lowered by 25 mm. As before, the car came only with an automatic transmission. The reason was not that Daimler-Benz did not want to offer the new manual transmission to its clients, the reason was that they did not have a manual transmission that they thought could cope with the car's torque for an extended period of time. The newly developed 7G Tronic transmission came with steering wheel gearshift buttons and allowed a wider spread of gear ratios and reduced the differences in engine speed between the individual gears. This way, the driver always had the optimum ratio available at almost any given speed. As already mentioned earlier, the 7G Tronic can skip up to four gears when shifting up or down, depending on the individual driving situation,

The AMG Speedshift had been developed to handle the additional torque and power. The torque converter lock-up from first gear optimized driving dynamics and increased efficiency by avoiding power losses in the converter.

Other Speedshift benefits were its ability to adapt to either a more engaged or more relaxed driving style. And using the gear lever's (or steering wheel buttons) touch-shift function, the driver could opt for either shifting automatically or manually.

When the gearbox' Sport mode was chosen, gears would shift with Speedshift some 35 percent faster than when in Comfort mode. If the Manual mode was selected, the car would stay in the gear selected and would not shift down, when for example under full load or kick-down. This way the driver with more than the usual sporting ambitions could fine-tune the car's huge performance potential to his liking. He did not have to, the automatic was superb, but he knew he could do, if he wanted to.

Details of the new 7G Tronic transmission that AMG further refined

The car rode on special 7.5 x 18 AMG alloys with 225/40 ZR18 tires at the front and 8.5 x 18 alloys with 245/35 ZR18 tires at the rear. One of the reasons for the 18 inch tires was brake clearance for the larger brakes. In order to handle all that power, the high-performance brakes offered composite brakes with floating, internally ventilated and perforated grey cast iron disks at the front. Maximum heat resistance was ensured for the 340 mm disks through an aluminum bowl and six-piston calipers with large surfaces of their linings. The rear came equipped with four piston fixed calipers and 330 mm solid brake disks. Other features such as Brake Assist, ASR and ESP had all been adjusted for the extra speed potential. It

was not only the technical side that impressed, also the interior was well appointed with many items, such as heated Nappa leather seats or interior lighting package coming standard. It was available in four different colors and covered also the armrest and center panels of both doors.

In order to optimize lateral support through fast corners, the seats had more defined side bolsters and Alcantara inserts for additional grip in the seats' should areas. Further refinements included an AMG instrument cluster with a 320 km/h (for the US: 200 mph) scale and an AMG styled steering wheel.

It was clear that all this had to come at a price. And although a base price of €69,830 (incl. 19 percent VAT) was not considered a bargain, it was seen as fair, when one had a look at what the car had to offer. In the UK it started at £50,879.89, while it cost in the US $66,000.

The *UK Telegraph* had this to say about the car:

"The SLK 55 AMG is flabbergastingly fast. It is possible to find yourself doing 100mph in this car just by wiggling the toes of your right foot in your shoe at 70mph. If it weren't held back by electronic limiters, it could certainly top 200mph. 0-60mph acceleration is at the speed of the fastest thing on the road - the courier's motorbike. In the short time that the SLK 55 AMG was in my hands, I doubt if I explored more than half of its performance capability. You could live with this car for five years and, unless you took it to a circuit like Spa or Nürburgring, a third of its powers might remain hidden from your grasp. The £350,000 Mercedes SLR McLaren is not much faster in a straight line and it handles so piggishly that the SLK is probably quicker on real roads in what we lightly term real life."

Red calipers were optional

Alcantara was an AMG exclusive

Also *Road&Track* came away mightily impressed:

"*After driving the new Mercedes-Benz SLK350, I wrote in our June 2004 issue: "If the SLK350 is the best of Veuve Cliquot, then we can expect the V-8-powered SLK55 AMG to be Dom Perignon." Now I can tell you that I have had my Dom Perignon! I wondered how the SLK55 would cope with 355 bhp. Surely, in the French southern Alps where I was driving, if I floored the accelerator out of a sharp uphill bend in 1st or 2nd gear, the rear wheels would start spinning, but the electronic differential control sees to it that this does not occur prematurely. If the driver insists, the ESP intervenes to stabilize the car, but late enough to let the driver enjoy and control a moderate power slide. AMG's special springs, dampers and anti-roll bars and 18-in.-diameter wheels shod with high-performance 225/40ZR-18 front tires and 245/35ZR-18 rears ensure grip, a neutral and nicely torque-sensitive cornering attitude and very small roll angles.*"

The SLK55 AMG Black Series

Specialty cars were made by hand, as they usually came in limited numbers only. The SLK55 AMG Black series was such an attempt and Daimler-Benz expected to sell around 130 of these built-to-order cars. Again the question: were they needed? Again the answer: of course not. So why the bother? AMG wanted to demonstrate their capabilities. Naturally they knew of cars tuned by Brabus or Carlsson and naturally they were aware that Daimler-Benz board members occasionally ventured into Brabus territory to have their "standard" AMG company cars somewhat improved. This Black Series served as testament to what AMG could accomplish, if they were allowed to. They were expensive to build by hand, but it helped AMG engineers to gain experience for even grander projects in the future (the SL65 Black Series and SLS come to mind) and it could be expected that AMG did sell each one of them at a profit.

Affalterbach: A Black Series next to a Special Performance and F1 Safety Car

96

5.4 L V8, 400 hp and 520 NM (384 ft-lbs) of torque, available at 3,750 rpm, with 80 percent of it already available at 1,850 rpm. Is there any need to say anything else? Yes, one more thing: the Black Series is despite its low slung, wider appearance and its beefier tires some 47 kg (103 lbs) lighter than the standard AMG. All this meant that one hp had to carry just 3.8 kg or 8.4 lbs (SLK55: 4.3 kg or 9.5 lbs). However, it also meant that this car was not for the average sports car enthusiast. Although advertised as an everyday usable car, one single drive over poorly maintained roads made it pretty clear that it was not meant for that kind of life. On the contrary, packed with expensive technical features of a super-car, it was an irresistible offer for the true hardcore aficionado. It was made for the person, who cherished the feeling to master a vehicle on the racecourse that was basically a powerful engine on wheels with the cabin squeezed in as sound studio.

The car raced from zero to 100 km/h in just 4.5 seconds (SLK55: 4.9) and sped past the 200 km mark in a mere 15.5 seconds (SLK55: 17.5). Top speed was reached at 280 km/h (173 mph), but it could be assumed that the car was capable of reaching somewhat higher velocities, if allowed to. A rev limiter kicked in at 6,750 rpm.

Switch off ESP and you could smoke them

The increased power of the engine came from modified air intakes, reduced resistance air filters, a new valve-timing scheme, and a revised exhaust system. In addition, the electronic engine management had been reprogrammed.

Naturally the car was also otherwise heavily altered, no wonder it had to be built by hand. The first issue that had to be dealt with was the retractable roof, which was replaced with a weight-saving one-piece carbon fiber reinforced plastic fixed top. This alone made the car some

35 kg (77 lbs) lighter. Another added benefit was the lowered center of gravity.

Then the new car came with a captivating redesigned front apron with wider openings, as it needed to supply the additional transmission and steering oil-cooler with plenty of air. New black carbon fiber side outlets helped to get rid of that air again. Wider front wings (plus 25 mm at the front and 10 mm at the rear) were made of reinforced Pu-Rim plastic, with carbon fiber composite materials added for additional strength.

The special Pirelli Corsa tires helped to save additional six kg

Another striking feature were the large 19 in multi-spoke forged alloy wheels with tire sizes 235/35 ZR19 at the front and 265/30 ZR19 at the rear. They filled out the wheelhouses perfectly and came with special Pirelli P Zero Corsa tires (Pirelli P Zero Nero was the second option), which delighted with superb grip even under adverse conditions. This set-up was some 6 kg (13.2 lbs) lighter than the standard SLK55 equipment. Adequate braking was ensured through high-performance brakes with large composite 360 x 32 mm disks and six-piston fixed calipers at the front.

The 330 mm internally ventilated and perforated brake disks at the rear were not changed. The adjustable sports suspension was optimized for the race track with tunable shock absorbers to adjust damping comfort and driving behavior. Ride height could be either lowered or raised by 5 mm and an additional brace across the engine compartment ensured additional body rigidity in fast corners. The speed-sensitive power steering came with a larger servo pump to make steering even more responsive under all driving conditions. Even sportier handling could be provided with a limited slip differential for €3,475.

The interior did not disappoint either. Newly developed sports bucket racing seats with no side airbags greeted the eager driver. They were covered standard with black pearl velour. If the new owner thought that a bit too frugal, he/she could always opt for the Leather Package, which included for €6,310 Nappa leather not only for the seats, but also for the dashboard, center console, door panels and rollover bars. Side airbags and leather in both doors were replaced with elegant looking carbon fiber elements with a superb finish. Carbon fiber covered also the upper part of the handbrake.

The sports steering wheel featured a black leather/Alcantara combination. If more carbon fiber was needed, the owner had two Carbon Fiber Packages as options. The first one cost €2,630 and was used in the interior, where additional carbon fiber was applied on instrument cluster cover, steering wheel multi-function button covers and gearshift lever. The second package addressed the car's exterior and cost €4,325. It came with additional carbon fiber for the front grill inserts, external mirror covers and the trunk spoiler lid.

Black Series lettering could be found on the carbon fiber trim above the glove compartment and at the AMG badge at the trunk lid. The first models were delivered in July 2006 with production coming to an end in March 2008. It was never officially offered in North America and cost €92,500 without taxes.

Carbon fiber at the doors meant that the side air bags had to go, which meant no registration possible in the US

First upgrades

General, early 2006:

The R171 received two years prior to the facelift a few updates. All plastic interior materials received a slightly softer, easier to clean surface and the AMG Sport Package came now also with the SLK55 spoiler lip at the trunk.

General, mid 2006:

The remote key came with chrome applications at its sides and the steering wheel shift buttons were changed to shift paddles behind the steering wheel. The code 428 remained unchanged. For some reason it could not be combined with a heated, leather/wood or wood steering wheel. Leather color "universum-blue" was taken off the options list.

General, autumn 2006:

Standard and Nappa leather had a softer structure and black Nappa leather became available for the SLK200 and SLK350. It meant that the codes for leather interior were changed.

SLK55 AMG Performance Package:

The AMG owner, who was a bit into racing, could order from August 2006 onwards a Performance Package. It was developed by the newly opened AMG

Performance Studio for three Mercedes models: the SLK55, SL55 and CLS63. It included various chassis and brake system modifications, but no alterations to the engine. Although this package was developed in Affalterbach, the car equipped with such package was built at the production plant in Bremen.

The internally ventilated and perforated composite front disks with six-piston fixed calipers were enlarged from 340 x 32 mm to 360 x 32 mm. At the rear, internally ventilated and perforated brakes offered 330 x 26 mm disks with four-piston fixed calipers. The 18-inch multi-piece alloys came with a Performance package specific design, called AMG Styling IV. Tire-size did not change though.

The suspension was further upgraded and stiffened, so that any driver with racing ambitions could try the car's potential on his/her favorite track. The interior offered carbon fiber trim and an AMG sports-steering wheel with a flattened bottom. Covered with black leather/Alcantara it was also smaller in diameter and came with aluminum shift paddles.

Shock absorbers were adjustable and six piston caliper brakes at front were enlarged to 360x32 mm

With the exception of the composite brakes, all other components could also be ordered individually for the SLK55. The package was available at a surcharge of €5,220 incl. 16 percent VAT.

For a limited time the Performance Package could be further upgraded with an engine uning that had already propelled the Black Series and lifted total power output to 400 hp. At the cost of €11,400, plus a further €1,400 for installation, it was a rather expensive proposition, but that way you had the Black Series car in the SLK package, meaning you could make full use of the vario-roof.

AMG Sport Package for SLK200, SLK300, SLK350:

In 2006, the trunk lid received the small spoiler that was previously SLK55 exclusive. North American bound vehicles always had the spoiler lip installed. Everything else remained unchanged.

Exterior paints:

During a car's production cycle all automotive manufacturers change their paint choices. Daimler-Benz is no different, so in June 2006 various color options such as *thulit-red* or *calcit-white* replaced *amber-red* and *alabaster-white*. Some options such as *designo varicolor IV grey-blue* were dropped. The potential owner could choose between three standard colors, eight metallic colors and one special metallic color, called *prenith-green* (option code 430). On top of that were various designo paints. When browsing through the SLK catalogue, you will eventually end at which exterior paints should go with which interior colors. In most cases it would be differentiated between 1. Possible and 2. Not possible. With the SLK, there were three options: 1. Recommended, 2. Possible and 3. Not recommended. It did not say "not possible", it said "not recommended". Did it mean that you could still opt for it, if you wanted to? If not, it was at least a nice way of saying "No".

SLK 280 in a rare designo Chablis metallic (043)

The 2008 facelift

Body and interior

For most car manufacturers it was almost mandatory to have a light upgrade of their models around halfway into their life cycle. Also Daimler-Benz offered for most of their vehicles a so-called mid-term facelift. Just in time for the 2007/2008 holidays, Daimler-Benz showed the "new" SLK with some 650 newly developed parts, to the press. The public would see it first at the January 2008 Detroit Motor Show. It should become available in April 2008. After the small sports car/grand tourer had managed to sell some 500,000 units in the first ten years of its existence, these changes should ensure that the concept would remain the most popular vehicle in its market segment.

Changes to the exterior were an arrow-shaped front apron with modified air-dam. The grill area around the Mercedes star was more enhanced and the fog lights came now with an elegant looking chrome ring. The rearview mirrors were slightly enlarged and featured LED indicators in a new design. The car's optic was further upgraded by a wide selection of new alloys. The rear apron offered a diffuser-style lower section. Its aerodynamic benefits might be questionable, but at least it looked good. This was enhanced by SLK55-style darkened taillights and new trapezoidal exhaust tailpipes with chrome tips. The rear antenna was slightly shortened to make it more suited to the forces of automatic car washes.

A natural beige interior with burr walnut wood

Many of the interior details were optimized with better material selections. The most obvious change was a redesigned three-spoke sports steering wheel with multifunction buttons and a new instrument cluster. It came with red needles and bezels in 3D-look, a fuel and a clock gauge and chrome rings for the instrument tubes.

The gear lever came with new chrome and leather trim and the power window switches were redesigned. Also a new "Gullwing red" interior color was added to the list of appointments. Although the SLK had nothing to do with the iconic 1950s sports car, the color was a close match to a color option once available for the

300SL and should help to shed some glamour on the SLK. Another new color scheme was "Natural beige", which complemented two equally new wood trim options: "Pale Burr Walnut" and "Black Ash Grain".

The COMAND APS was upgraded from version 1 to version 2.5 and is already described in the earlier COMAND chapter in this book. Standard features in every radio offered by Daimler-Benz was a hands-free operation with Bluetooth technology plus a new media interface in the glove compartment. It allowed full integration of most mobile devices. They were fully controllable with the audio system's user interface.

New (above, here a UK SLK) vs. old

The more aggressive look of the new arrow-shaped front apron is immediately noticeable

The new steering wheel with its revised multifunction buttons

The revised engines

Changes to the exterior and interior were welcome news of course, but would not have really helped to keep the emotions and interest in the car high. So the best news came from the engine bay. Daimler-Benz engineers did not only manage to make the engines (except the SLK55 AMG) more powerful; they also made them more frugal in their fuel consumption, thus improving their carbon-dioxide emissions.

The SLK200 Kompressor (new internal code: 171.445) offered a modified engine management system, a fine-tuned supercharger and upgraded pistons. That way it produced an extra 21 hp with a total of 184 hp, available as before at 5,500 rpm and had its torque increased from 240 NM to 250 NM (177 to 184.4 ft-lbs), available from 2,800 to 5,000 rpm. It accelerated from zero to 100 km/h in 7.6 seconds and could reach a top speed of 236 km/h. The astonishing thing was that fuel consumption could be reduced by a full 1 l to an average of 7.8 l / 100 km (30.1 mpg).

The SLK200 engine was for most people quite sufficient for their daily commute

110

The 3.0 L engine retained the cover with its grayish V

The SLK300 was improved with regards to fuel economy, its power output was left unchanged. Fuel consumption was reduced by 0.4 l to an average of 9.4 l / 100 km (25 mpg).

Thanks to a new engine management and a host of other improvements, the SLK350 (new internal code: 171.458) came along with a 33 hp beefier engine, which made total power output come to 305 hp, available at 6.500 rpm. But in true sports car fashion, the rev limiter only kicked in at a very respectable 7,200 rpm. The optimized pistons with a curved surface allowed for a slightly raised compression

ratio (1:11.7 instead of 1:10.7). The pre-facelift engine used flap-controlled intake manifolds, which came with the slight disadvantage of not being able to always ensure a high cylinder charge at high engine speeds. Tests with new single-stage intake manifolds had shown how to overcome this small drawback. Further tests had revealed that when made out of plastic, they would be even more efficient. The core melt-out technique could be applied on plastic, which gave the engineers a higher degree of flexibility than they previously had with die-cast versions.

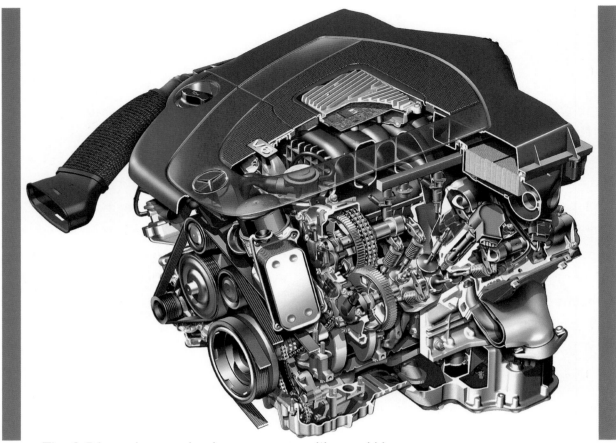

The 3.5 L engine received a new cover with a red V

Another advantage that came with the use of plastic was its limited heat conductivity, which helped the intake manifold to stay cooler. This in turn was beneficial for the cylinder charge.

The single-stage intake manifold solved the issue with continuous high charge at engine speeds above 4,000 rpm, but it came with a drawback, as it was less efficient at lower rpm. After more testing this was finally solved with the camshaft position advanced by five degrees of crank angle on both the intake and exhaust sides, the high charge could now be also guaranteed under partial load or when at idle. Higher engine speeds forced the engineers to look at the valves, which needed to be modified.

New lightweight valves were made of high-strength steel that was more heat resistant and new sodium-cooled valve stems for the exhaust valves offered now a bore of 3.4 mm. Another potential problem with high revving engines was to make sure that the valves were reliably closed at all times.

With conventional cylindrical valve springs, oscillations could build-up over time, so that they could not close the valves properly, once higher engine speeds were reached. Conical valve springs on the other hand offered heavily damped natural vibration characteristics and were thus capable of closing the valves also at high speeds reliably. Problem was that conical valve springs were longer. At one point it seemed that trying to improve one issue, immediately opened up a stream of new tasks.

In order to make the new valves fit, the intake port for optimum cylinder charging needed to be changed. Also modifications to the cooling jacket were required, so at the end the engineers decided that it was best to design a completely new cylinder-head casting.

Another problem that could occur with engines that run for an extended period of time at speeds higher than 6.300 rpm, was chain-drive vibrations. Ultimately, these vibrations could shorten the chain-drive's life. In order to counter-act the forces acting on the chain and thus the vibrations, a new tri-oval chain sprocket was installed, its triangular form barely noticeable. Its tri-oval shape generated inversely-phased amplitudes, which in turn limited chain vibrations. In order to make room for these changes, the vibration damper on the belt-pulley needed to be adapted.

One factor that was always liked with the SLK350, was the sound it produced. Over the years sound engineering had almost become an art form, as car manufacturers came to realize that there was an emotional side to it that customers wanted the companies to address.

So for the facelift, Daimler-Benz engineers had another look at the car's exhaust system, but decided at the end not to change anything. Instead, they turned to the air cleaner and modified that one. The effect was a wonderful throaty sound, which the driver could enjoy, when the car was decelerating.

It was especially noticeable with the automatic cars, because in combination with the 7G-Tronic automatic, the engine management system blipped or double-declutched the throttle automatically during downshifts. This had the additional benefit that it reduced the load alteration effects, as it equalized the rotational speeds of the crankshaft and trans-mission.

33 percent of all R171 produced were silver, making it the most popular color

All these measures helped the SLK350 to accelerate from zero to 100 km/h in 5.4 seconds and reach an electronically limited top speed of 250 km/h. Maximum torque had been slightly upgraded by 10 NM to 360 NM (266.5 ft-lbs), of which 95 percent were already available at just 2,000 rpm. These figures are superb in themselves, but they tell the interested reader very little about the almost prodigious pulling power and the tremendous fun a manual shifter could give its excited driver. How the car pushed from almost standstill to 150 km/h, racing through its gears with verve and delightful

elasticity needed to be experienced. And one had to remember, this was not a car with an engine carefully hand-assembled by a single gifted technician in Affalterbach, this was a car with an engine that came off the huge Daimler-Benz assembly line. Surprisingly all this did not come with a big penalty at the petrol station. With an average consumption of 9.7 l / 100 km (24.2 mpg), the car was just a bit thirstier than the SLK300. In order to keep fuel consumption in check, one measure was a slightly longer rear axle ratio. It enabled the car to pass the 100 km/h mark in second gear.

114

An issue that became increasingly more important with modern cars was their carbon dioxide emission. Large cities started to penalize owners of high emitters. One such city was London, where in the 2008 Mayoral elections the two contenders Boris Johnson and Ken Livingston laid out plans how to hike daily rates for cars entering London.

Contrary to Johnson, Livingston threatened to increase daily rates for cars emitting more than 225 g/km from £8 to £25. Cars that emitted less than 226 g/km would stay at £8. However, it were not only large cities that started to take on high emitters, it was also the government that started to look at such issues. In the UK for example, cars with emissions exceeding 225 g/km would have their Vehicle Excise Duty (VED) increased from spring 2009 onwards to £415 annually, for those emitting more than 255 g/km to £440. And from spring 2010 onwards, additional purchase taxes for those cars of £750 and £950 respectively would have to be paid.

So for Daimler-Benz it was not only important to have the overall fuel consumption reduced, it was equally important to have a check on emissions. It was a small wonder that they managed that all three SLK fell below the 225 g/km threshold.

Obsidian black came in second with 26 percent

115

The manual SLK200 had its emissions reduced by 25 g to 185 g/km, the 5-speed automatic version by 19 g to 192 g/km. The manual SLK300 had its emissions reduced by 11 g to 220 g/km, the modern 7G Tronic automatic version by 6 g to 216 g/km. Even the SLK350 was able to reduce its emissions for the 7G Tronic version by 17 g to an almost magical 219 g/km (for such an engine size). The manual car missed it by an inch with 227 g/km. Naturally the SLK55 could not live up to such values, its output remained before and after the facelift at 288 g/km.

A feature that was little understood, when introduced to the SLK, was the Direct Steer System, previously called Speed-Sensitive Steering System. It came standard with the SLK55 and was optional for all other SLKs. Contrary to its predecessor it acted purely mechanically, doing away with elaborate electronically governed actuators and complex sensors. It all started with a new rack with slick gearing, making sure that the steering gear ratio changed in tune with the actual steering angle. That was nothing new of course. For an excellent straight-line stability especially at higher speeds, the ratio was indirect with the steering wheel around the center position. However, when deviating from this position by just five degrees, the ratio started to increase fairly rapidly with

the result that steering felt much more direct and responsive.

With the new system fitted, the turns needed lock-to-lock were reduced by some 25 percent, which meant that even in city traffic, very little steering was required to change the car's direction. This was most noticeable on country roads, where the fun factor especially at higher speeds was greatly increased with an amount of precision and agility, not often seen on other cars. But it also meant that coming from another vehicle, it took some getting used to, before the much more precise steering could be fully appreciated.

In Germany, the revised SLK200 Kompressor started at €36,503, while the SLK300 cost €41,858. The SLK350 started at €46,980 and the SLK55 AMG at €69,050 (all prices incl. 19 percent VAT). In the US, the SLK300 was offered at $44,150, the SLK350 at $49,450 and the SLK55 AMG at $63,200.

In the UK, the on the road price for automatic cars, including 17.5 percent VAT, delivery charges, number plates and VED was for the SLK200 £29,750, the SLK300 £33,010 and the SLK350 £36,960. The SLK55 AMG started at £50,760.

A BMW Z4 3.0Si would cost in the US in comparison $42,700, while the M Roadster was available at $52,400. A Porsche Boxster cost $45,800 and a Boxster S $55,700. Especially the Boxster S price appeared like a bargain, when compared with the SLK55. But of course, it lacked the V8 and the vario-roof.

Changes for 2011 models

Production of the R171 ended in December 2010 to make room for the new R172, of which production started in June 2011. In order to accommodate the necessary tooling changes and to slowly reduce and finally stop production of the R171, the last year production units were introduced with fewer options in March 2010 as 2011 models. The designo program was not anymore available and the only leather option was black. The SLK55 continued until its end as 2010 model. Until production was stopped, all options, including the four Napa leather color variations were available to customers. Its only new standard feature was a glass wind reflector instead of the fabric one.

These last model year cars came as Avantgarde Edition and were equipped with black roof liner, AirScarf and heated leather seats. The SLK350 offered standard power seats with memory package, power steering column and heated steering wheel (manual car only). Another SLK350 standard equipment was the Harman/Kardon Logic 7 audio system plus in the US Sirius satellite radio.

Other standard goodies were borrowed from the AMG Sport Package, which was not available anymore: twin five-spoke 17 in alloys (SLK300, code 32R) or multi-spoke alloys (SLK350, code 796), tire rating, AMG trunk spoiler, carbon interior trim and smoked headlamp surrounds. The SLK55 owner could choose between two 18 in twin five-spoke alloys (codes 795 and 786).

The special editions

By 2006, when the first special edition R171 was introduced in form of the *"Edition 10"* (code P38), special editions were nothing new anymore for Daimler-Benz. This was a company that for many years did not see any benefit in offering such models to their customers. And the customers did not seem to mind. A first such attempt was made in 1995 with the SL R129. With more competition available, sales had declined and suddenly special editions in small numbers seemed to make perfect sense to keep interest in the car high. At the end, marketing must have liked the idea a lot, because the R129 was offered with 17 special editions. All of these editions were mostly aesthetic tweaks though, the engines remained untouched.

2006 was of course a great year for a special edition celebration, after all the small sports roadster had celebrated ten most successful years in the market and had achieved to surpass even the highest expectations of its maker. No wonder, the first such edition was called "Edition 10".

"Edition 10" was launched at the "Mondial de l'automobile" fair in Paris in September 2006 in a matt-grey metallic paint (magno allanite-grey) that set it apart from the standard SLK and should demonstrate a future color-trend of Mercedes motorcars. Other features that made it special were grey-metallic painted ten-spoke alloys embellished with black-and-silver Mercedes star emblems, 7G-Tronic, Parktronic plus darkened taillights. The perforated, heated black leather seats were electrically adjustable with "Edition 10" logo at their headrests. They came with silver-colored underlays and silver top-stitching around the seats edges. The liner of the roof and rear window surround were an all-black affair with a black trim strip on the driver's side of the steering wheel and a lacquered metallic trim piece above the glove compartment. Velour floor mats had black linings and embroidered "Edition 10" logos, which were also featured at the front fenders. In order to emphasize the car's sporty character, the speedo and rev counter both offered great looking red scales.

The 231 hp V6 engine in the Paris show car came from the SLK300. Standard options included AirScarf, Thermotronic air-conditioning, the COMAND system with Europe-wide navigation and MP3-capable CD-changer and at the front darkened headlight inlays.

It became available in January 2007 in a limited edition of 350 units with black (code 040) as standard finish. The "Edition 10" model was available for all SLK versions, except the SLK 55. Depending on the model a surcharge between €1,785 (SLK200) and €1,428 (SLK350) had to be paid. Next to the matt finish (code 044 at €2,498) as shown in Paris, also obsidian black metallic (code 197) and iridium-silver metallic (code 775) were color options at €684. But all the expensive goodies of the Paris show car such as AirScarf, Thermotronic and Comand system had to be ordered separately though. They were not part of the official prices, which only included the standard paint, wheels and interior trimmings.

An "Edition 10" model (here without badge) in a rare matte green

The second special edition was the *"Sport Edition"* (code P84), launched in September 2007 and available until November 2007. As before it could be ordered with all engines except the SLK55 AMG at a surcharge of €3,950. It offered basically the same exterior and black interior refinements with red top-stitching that the later introduced Sport Package, which has been discussed already, offered. But it also included for the SLK200 air-conditioning and for all three cars the fine AirScarf feature. The car came standard in iridium silver, other colors were also available. In case a designo scheme was ordered, the red stitching would be replaced with a stitching fitting the respective designo interior.

Never short of eye-catching ideas, the sales and marketing department initiated for the 2009 Geneva auto show another special edition, the *"2Look Edition"*. At a surcharge of €2,100 it was available for all SLK models, except again the SLK55 and would be limited to 300 units. It was developed with a focus on younger drivers under the theme: *"the attraction of opposites"*. Consequently it came in just two exterior color schemes: black and calcite white, which the marketing people called in their ad "glossy black" and "lustrous white". At a surcharge of €580 it could be ordered in Obsidian black metallic (code 197) and for €1,990 a special finish "designo mystic white 2" (code 048) was available.

A "2Look Edition" SLK in designo mystic white 2

Both black and white "2Look Edition" cars had the same interior color scheme

The black and white theme continued inside the car with white, heated Nappa leather on the seats' center panels and black Nappa leather on the contoured bolsters. Also the door panels and center armrest were covered accordingly. All this was topped by some great looking white decorative seam, white-painted trim on the otherwise black dashboard and a black roof lining. 2Look Edition labels could be found on the white-edged floor mats, the acrylic glass draught stop and the sides of both front fenders.

The car rode on newly developed 18 in twin five-spoke alloys with the standard SLK350 tire-size. The alloys were offered in two different surface variants: high-sheen finish in titanium silver, which was usually used for the black car, and a darker high sheen version with so-called chrome shadow finish, which was normally reserved for the white car.

In early 2010, the **"Grand Edition"** (code P61) was launched together with the SL R230 "Night Edition". Both cars did not have any production limits, but it can be assumed that in case of the SLK, not more than 300 units had been sold. The designer's motto was: "masterpieces that dreams are made of".

At surcharges between €3,810 (SLK200) and €2,740 (SLK350) this last R171 edition could be ordered in almost any SLK exterior color and the special designo color graphite metallic at €1,800, which looked especially fitting with the newly designed 18 in triple five-spoke alloys in their glossy sterling silver finish.

A sport package was not part of the equipment, but could be ordered separately. Silver fins, color matched headlamp housings, a darkened third brake light and Grand Edition emblems at both sides of the front fenders complimented the car's elegant appearance. The interior did away with the usual black in favor of a restrained, stylish basalt grey with red top-stitching.

The heated seats were redesigned and used around the bolsters a slightly lighter designo grey leather with red top-stitching. The same material covered also the door armrests. Basalt grey was used again for the door panels, the seat belts and as border for the floor mats, which also carried the Grand Edition logo. The aluminum trim parts were replaced with leather ones in basalt grey and Grand Edition logo.

The "Grand Edition" SLK. Many of these special edition cars were sold as SLK200

The "Grand Edition" came with a unique seat cover design

The cars came again with AirScarf and for the SLK200 additional air-conditioning, which explained its higher price. If one would buy these goodies separately for example for the SLK200, one would have to pay (with designo leather) around €8,300, which should demonstrate that the Grand Edition was excellent value for its money.

2010 saw another unique SLK unveiled. It was called *"SLK200 Naked".* You don't believe me? Please read on. It was produced as a limited edition of just 99 units for Italy, hence the choosing of the smallest engine. Cars with a displacement higher than 2.0 L are higher taxed in Italy. That is why the SLK200 was the most popular model there.

It was launched at the World Ducati Week from June 10th to June 13th 2010 and sold between June of that year and January 2011 at €40,900. It was aimed at a younger clientele and marketing was pushing the nature, outdoors and sport idea. Whether the name lived up to that idea is anyone's guess.

Besides the somewhat unusual name, the car did not offer anything off the beaten track though. It came with the usual goodies that had been offered on previous special editions such as air-conditioning, Audio 20 system, heated Nappa leather and aluminum door entrance inserts (but these did not carry the name "Naked"). The car came further with 18-inch twin five-spoke alloys, headlamps with darkened surroundings and a spoiler for the trunk lid, where it was greeted at the right side with the "Naked" logo. It was offered in four color options: limestone white, palladio silver, obsidian black and palladio grey. On special request and at a surcharge, also silk cashmere white and alani platinum grey could be ordered.

Sorry for the less than perfect quality, but photos of a "Naked SLK" are hard to get

The SLK 55 AMG Asia Cup

A proof that Daimler-Benz was to a certain extent willing to listen to customer requests can be seen with the SLK55, which was on behalf of an Asian company, called *"Front Row",* modified for their ten "Asia Pacific Cup" racing activities in China, Indonesia and Thailand. They were named SLK55 Tracksport and came in two versions, one was a race car and a slightly altered one was a street car that should be used for various pan-Asian lifestyle and driving events. 35 of these rhd cars were built at Affalterbach in summer 2006 and cost some €100,000 each. This put them between the Porsche Carrera S and the GT3.

The Tracksport race car came with the standard power output of the SLK55, which meant its 360 hp was available at 5,500 rpm, while its maximum torque of 530 NM (391 ft-lbs) was delivered in a wide band from 2,800 to 5,400 rpm. The Tracksport race version accelerated from 0 to 100 km/h in 4.7 seconds, reached 200 km/h (124 mph) in 17.3 seconds and achieved a top speed of 280 km/h. The Tracksport street version had its exhaust system optimized and offered 400 hp instead. That meant it accelerated in 4.5 seconds to 100 km/h and was capable of reaching 300 km/h.

Cockpit of the Tracksport

The same set-up was used for the AMG Performance package, offered for the standard SLK55. Both Asia Cup cars had a gear ratio of 1:3.27 and a rear axle cooler that was borrowed from the SLR McLaren. The racecar version did not have the 5-speed automatic. Instead, it used the manual mode of that transmission with two options. The M1 mode offered slower, more comfort-oriented gear-changes (who wants that in a racecar), while M2 gave the driver an instant shift of gears.

In order to have the car's weight optimized, AMG followed more or less the path already taken with the Black Series and the Formula One Pace car: they replaced the vario-roof with a fixed version made out of carbon fiber reinforced plastic. That saved some 75 kg.

Also the passenger seat was removed for the race version; another 25 kg saved. The steel fenders were replaced with slightly wider carbon fiber composite versions. Air-con and power windows were kept though. Asia is usually hot and humid.

But at the end the car still had a curb weight of 1,469 kg (3,232 lbs), which was explained by AMG project manager Hendrik Hummel as follows: "We had to use larger alloys and tires and also the brake system was taken from the AMG Performance Package. Then we needed additional oil-coolers for engine, power steering and gearbox. On top of it comes the not really light roll-over cage". The street version had a few more changes, as it came with even wider fenders, two lightweight sports seats with conventional seatbelts and an adjustable suspension.

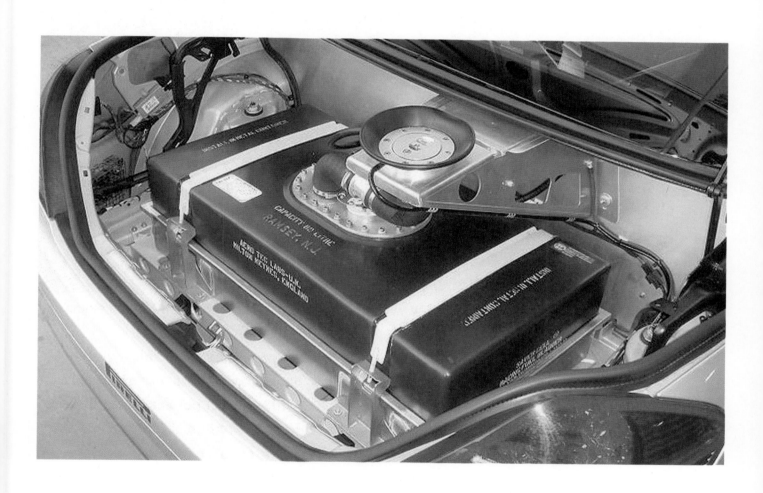

The SLK320 CDI Tri-Turbo

Daimler-Benz showed at the March 2005 Geneva Automobile Show a couple of interesting cars, all equipped with newly developed diesel engines. This way the company wanted to demonstrate that it was the first automotive manufacturer worldwide to equip all its models with particle diesel engine filters as standard. For demonstration reasons this also included the SL400 CDI and the SLK320 CDI, both of which would not be made available to the public. Diesel development had come a long way with Daimler-Benz, who introduced their first diesel car in form of the 260D in 1936. The engine in the 2005 SLK would use one tenth of the diesel fuel that the 1936 car would require for touring the same distance. And when compared with a 1995 Mercedes diesel, it had its carbon dioxide emission reduced by 85 percent,

The SLK tri-turbo engine was based on the new V6 diesel technology with a two-stage charging system. It consisted of three turbochargers, two of them positioned next to the cylinder-banks, while the third one, which was larger in dimension, was positioned between the V of the cylinders. At lower loads, air would go through all three chargers, with the two smaller ones doing most of the work. With engine speed increasing, more air would pass through the larger turbocharger and the two smaller ones would be gradually bypassed.

At full speed only the third charger would operate. This way it was ensured that full power was available at almost any engine rpm. In order to increase performance figures further, the intercooler and pipe diameter (for charge air and exhaust) were enlarged.

All this resulted in 286 hp and an impressive 630 NM (464.7 ft-lbs) amount of torque, of which 90 percent was already available at 1,500 rpm. The diesel-powered SLK accelerated from zero to 100 km/h in 5.3 seconds, travelled a distance of 1 km in just 24.4 seconds and had an average fuel consumption of 7.5 l diesel (31.4 mpg). Still, despite all these impressive data, Daimler-Benz was hesitant to offer a diesel in a sports roadster. That would only change with the R172.

Experiencing the SLK55 AMG

In my book about the R170, I drove the SLK230 Kompressor and in the book about the R172 I tried an SLK250 CDI. I hope you will forgive me that his time I indulge in something a bit different.

If you were in let us say 2005/2006 in the market for a high-end, high-performance small open sports car, choices were few and it did not really help to look around in your country club's parking lot, because most probably you will see there a line of pretty Porsche Boxsters with the occasional BMW M roadster dotted in-between. The SLK did not really spring to one's mind in those days, when it came to high performance, which was probably due to the first SLK's image as a slightly feminine looking, cute softy. Daimler-Benz had tried to remedy this with a more aggressive look of the second-generation car. But still, image-wise it was not there yet, which was a shame, because both the SLK350 and especially the SLK55 AMG were and still are brilliant performers and deserve a much closer look. Ok, the AMG is stuck with an automatic transmission. Despite its SpeedShift 7G Tronic with fully adaptive program for comfort and sport modes, Porsche's dual-clutch PDK offers a better job in a car that costs almost $10,000 less.

And here ends the (minor) complaining, because once you have the chance to experience the SLK55, all of this changes. From the outside, it does not differ much from its tamer cousins and in case those are equipped with the AMG Sport Package, chances are you will not spot the real AMG at all. So why bother and not settle for the less expensive (I hesitate to use the word "cheap") version. Hold your horses, that is not what we came here for today.

Because it is that fantastically vulgar big V8 under the hood that makes you forget about its siblings or the slight disadvantages of the slush-box. Actually, it makes you forget about anything negative you might have heard about the car. Because this two-seater gives you the nicest plunge imaginable into the world of insanity from the skunk-works deep inside the sacred halls of AMG. Put your foot down, if you feel felonious, it is pure distilled essence in single malt driving pleasure at any given speed. The little brute jumps forward like an F22 Raptor, with its four exhausts blurring out an almost intoxicating, addictive sound.

The toughest part of the acceleration excitement is actually the very beginning of it. Because you need to modulate the throttle quite carefully in order not to provoke a rear wheel spin. But once moving at more than 20 km/h, just put the pedal to the metal, take a deep breath and just hold on.

The engine will shoot to 6,700 rpm and shortly before the rev limiter kicks in, the 7G Tronic will shift, dropping the revs down to around 5,000 rpm. Then the rev climbing will start again. With your heart beating at an alarming rate and your blood pressure at stratospheric levels, you will experience this as an almost constant pull with just a few minor interruptions, when the transmission has shifted. And during the whole process you will hear the engine roar at different tempi. This is no mere *allegro assai*, this is absolute fantastic *andante furioso*.

Ok, the ride is firm, some would even call it harsh, but what do you expect from a car that corners like it is on rails. The seats actually do not look that comfortable, but they are. And they are well-shaped to help you survive those endless drifts you can provoke with the broadest possible grin on your face after you have switched off ESP. Keep it switched off, floor the pedal out of a sharp uphill bend in first or second gear and let the rear wheels spin and burn expensive rubber.

However, with ESP engaged, the special springs, dampers and anti-roll bars all ensure that you can enjoy and control a moderate power slide or nice, precise cornering with just minor roll angles. Cracks in the pavement that you see disappearing under the car's hood get instantly interpreted by your steering wheel. You know you are in command, so command.

After a few jaw-dropping, fast turns you finally need to stop on the side of the road. Because you want to relive for a few seconds in your mind the almost crazy excitement, you felt by pushing yourself (not the car) to the limit. Then you climb

out of it, slightly shaken but with raised eyebrows and silent nods. Yap, the bloody thing really is that good.

You can run the quarter mile in 12.5 seconds at 175.7 km/h (109.2 mph) and put the car repeatedly to a full stop from 100 km/h in 34.4 m (113 ft), all with an average fuel consumption of 13.7 l/100 km (17.2 mpg). For taller drivers like me it can be a bit tricky to get behind the wheel, but once properly settled in, there are no more complaints about lack of legroom. It all fits precisely. In the beginning you had heard me mentioning the 7G Tronic as being a bit outdated in comparison with the Porsche derivative.

In real life, you don't feel it as a disadvantage, because this fantastic engine covers everything. And because of this, you are best off, leaving these nice looking aluminum gear shift paddles alone.

The instrument cluster with its big gauges and red needles is easy to read. The speedo comes with a 320 km/h scale (in the US: 200 mph). Does it run that fast? Nope, but you almost wish it could. With the gear in "M" mode, the main menu displays the currently selected gear and gives up-shift recommendations. Climate controls are easy to navigate with twin rotary knobs controlling fan speed and airflow, while well-labeled buttons take

care of the rest. The COMAND's four-way keypad can be tricky to operate, there are much more intuitive systems around from Lexus or Audi. But as my car is equipped with Linguatronic voice control, all this is forgotten. In short, it works brilliantly. Just pull the dedicated stalk on the steering column and a list of voice commands are at your disposal. Say one of them aloud and the system does comply. As almost any common function is covered, just forget messing with all the knobs and buttons, just talk to your car. That is especially fun in summer with the top down at a red light and you are alone in your car. People stare at you and think you've gone mad.

With its relative light weight and all its power, the car is not that much fun in the wet or in winter, with one exception: drive it on a winter day in beautiful sunshine with the top down on snow and have the ESP switched off. Try it, it will beat anything you have experienced before. Lower the windows, switch off the radio and inhale everything nature and the sound of the engine throw at you.

The SLK55 is the smaller brother of the SL63 AMG, which cost almost double. And with a few extras like Harmon/Kardon etc added, the SLK55 ends up to be a car with a sticker price in excess of €75,000. Yes, it is expensive, but to be honest, it is at the same time still somewhat reasonable for a well-built Mercedes roadster with a wickedly powerful V8, which offers a full-strength dose of sports car charisma.

I have never seen a yellow SLK55, so if you allow me, I share these two photos with you

The sales performance

The first generation SLK sold from 1996 until 2004 some 311,222 units. The SLK230 was with 160,825 units the most popular. That was difficult, if not impossible to match. The R170 was at its launch a novel idea with its vario-roof. This concept made the small roadster a trendsetter. In 2004, when the R171 was introduced, that novelty had found numerous followers and had faded somewhat. Yet, although the R171 managed to sell "only" 242,184 units, it was at the end the more successful of both cars. Why? The R170 was produced for nine years, which makes for an average annual production volume of 34,580 units. The R172 was produced for seven years (counting for both cars the early 1995/2003 pre-production units into the first full production year), which accounts for an average annual production of 34,597 units, if one can compare it that way. It managed to beat its predecessor on an average annual sales level by exactly seventeen cars.

With the SLK in the Swiss Alps

It was no surprise that also the four-cylinder R171 proved to be the most popular model, as it sold 140,094 units. With 163 hp the SLK200 was sufficiently fast for most drivers. Its sales account for some 58 percent of total R171 production. Considering that the car was not sold in the US, it makes for the rest of the world an even greater share. It also tells quite a bit of a story of the US market share in total SLK sales, previously always the #1 market for anything convertible-like from Daimler-Benz. The SLK200 success also proves again that whatever journalists might think of a certain car with a so-called underpowered engine, it is not what people tend to follow. At the end, it is their wallet that makes the final decision and one can believe that the majority was mighty happy with their choice.

2004 was the best year for the car, as 34,378 SLK200 and 17,038 SLK350 could be sold. Even the SLK55 made a nice entrance into the market with 1,036 units for 2004 and 3,904 units the following year. From there it went quickly downhill for the fantastic car with at the end just 227 units in 2009 and 184 units sold in 2010, when production of the R171 was stopped. Nevertheless, a total output of 9,541 units compares quite favorably with 4,333 units of its predecessor (2001 to 2004).

From 2006 onwards, all SLK versions suffered. Only 30,859 units were sold in 2007. 8,638 units accounted for example for the SLK300, 4,215 for the SLK350 and 1,282 for the SLK55. The 2008 facelift failed to turn things around with a total of 29,853 units sold in that year. 6,237 units accounted for the SLK300, 4,679 for the SLK350 and 976 for the SLK55. 2010 marked the last year for the R171 with a total of 13,216 cars. The large majority of 8,285 units or 62 percent accounted for the SLK200.

Despite the facelift, the SLK was towards the end in its twilight years. It was offered to a market that is very fashion conscious and as soon as there are sexier contenders around, there is very little brand loyalty and people start to look elsewhere. Daimler-Benz of course knows this and had that is why shortened the production run to just seven years. One reason for the sales slowdown after just two years can be possibly attributed to the F1 inspired nose.

A complete sales breakdown by model and year can be found at the end of this book.

The Brabus SLK

Any Mercedes car is nowadays a subject for tuning. Even a stately Maybach, although no longer produced, is and has been heavily altered. So it should be no wonder that especially the SLK was embraced by the tuning scene, when it was launched in 1996. And this naturally continued with the second generation SLK. You may ask yourself, why on earth does anyone in his right mind want a car that has even more horsepower than the "standard" SLK55 AMG. After all the AMG throws already 360 hp at a curb weight of just a bit over 1.5 tons. Well, it is all a question of perception. Once you have driven the little AMG roadster for a few weeks, you might ask yourself indeed, whether there isn't anything faster available. Just for the pure fun of it of course.

The most popular tuner in Germany for anything Mercedes is next to AMG of course Brabus. And after AMG had been finally swallowed by Daimler-Chrysler in 1999, Brabus remained one of only two independent large European tuners (the other one is Carlsson) that concentrates solely on Daimler-Benz cars. Founded in 1977 by Klaus **Bra**ckmann and Bodo **Bus**chmann, it sells nowadays as a fully recognized automotive manufacturer around 8,000 cars annually. It was rumored that a few Daimler-Benz executive board members had their AMG based company cars further tuned by Brabus.

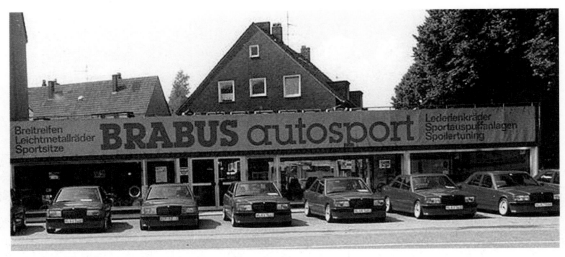

Humble beginnings: Brabus in its early days

140

One of the first R171 versions offered by **Brabus** was the SLK6.1S. It was pushed by Brabus' chief tech-honcho (and former Daimler-Benz engineer) Ulrich Gauffrés to 445 hp at 6,000 rpm. Brabus owner Bodo Buschmann calls him respectfully "the Professor". The car developed an impressive peak torque of 635 NM (468 ft-lbs), available at 3,800 rpm. Power was transferred to the rear wheels via a modified seven-speed automatic transmission and a bespoke locking differential. Equipped with the specially adapted, larger threw-flow Brabus exhaust, the car would race from zero to 100 km/h in just 4.2 seconds and reach an electronically limited top speed of 305 km/h (190 mph). For this the speedometer scale was changed to 330 km/h (for the US market: 205 mph).

The car came with bigger six-piston aluminum fixed caliper brakes with 355 mm vented and slotted discs at the front and 300 mm discs and four-piston aluminum fixed calipers at the rear. The suspension was stiffer and lowered by a full 30 mm and various Monoblock alloy sets were available from 17 to 19 inch. The largest tires that could be ordered were Yokohama or Pirelli 225/35 ZR19 tires at the front and 255/30 ZR19 ones at the rear. Did I say that the exhaust note was truly out of this world?

Like AMG Brabus puts a plate on each of its engines, but not with the name of the mechanic, who builds it, but with the name of the man, who designs it: Ulrich Gauffrés

But Brabus did not limit itself to the AMG car, it offered tuning versions and wind-tunnel tested body kits for all SLK models and started with a K4 tuning kit for the SLK200. It increased the pre-facelift car's output from 163 to 193 hp, with peak torque rising from 230 NM to 260 NM (170 to 192 ft-lbs), available from 2,500 to 4,800 rpm, just as the standard engine. While acceleration to 100 km/h is with 7.5 seconds just some 0.8 seconds better, top speed increased by 9.6 km/h (6 mph) to 246 km/h (153 mph).

Both six-cylinder cars could be ordered with a 4.0 L engine that offered an output of 332 hp at 6,100 rpm and a peak torque of 420 NM (310 ft-lbs) available from 2,800 rpm onwards. The 4.0 L car accelerated from 0 to 100 km/h in 4.9 seconds and reached a top speed of 281 km/h (175 mph).

The interior could be outfitted with almost anything the customer desired. Some of the more popular extras were soft Mastik-leather and polished stainless steel rollover bars. They would also come in a steel-carbon fiber combination.

Sinking money into a tuned SLK200 might sound like a silly idea to many readers, but in quite a few countries like the Mediterranean ones, or Hong Kong, Malaysia and Singapore for example, there is a heavy penalty to pay for larger displacement engines. For them a tuned-up smaller-sized engine does make indeed economic sense.

Mastik leather and a bespoke gearshift lever are very popular with Brabus customers

Other tuners

Another well-known tuner was **Piecha**. Towards the end of the R171 production run, they introduced the 2010 SLK Final Performance RS. It was developed in cooperation with Piecha's Swiss importer Auto Trachsler AG, used Piecha's best design programs and came in just two color options, both of which had never been officially available for the R171. This was either semi-matte fire-opal red or semi-matte diamond-white metallic with matching black Nappa leather with either red or white top-stitching. The distinctive front spoiler bumpers had integrated LED daytime running lights, wedge-shaped side skirts with air intakes. The rear bumper came with a wide grid core and special diffuser vane. The prominent rear spoiler lip was extended to the edges of the trunk.

The Piecha SLK in semi-matte diamond silver

144

The engine was left more or less un-touched, but came instead with a so-called Piecha power converter. That was not a chip-tuning feature and the power output remained unchanged. But Piecha claimed that it would boost power especially at lower rpm. Reactions to this "gimmick" have been mixed in Mercedes internet sites, most called it an expensive nonsense. While Piecha did not release any engine performance data, it equipped the car with a quadruple-flow sport exhaust, lowered H&R springs and 19 dp2 Phantom 8.5 alloys at the front and 20 in 9.5 alloys at the rear. Tire sizes were 225/35 R19 and 265/25 R20 respectively. As the body trim was custom-made, it could be fitted to any R171 and assem-bled in the color of the donor car.

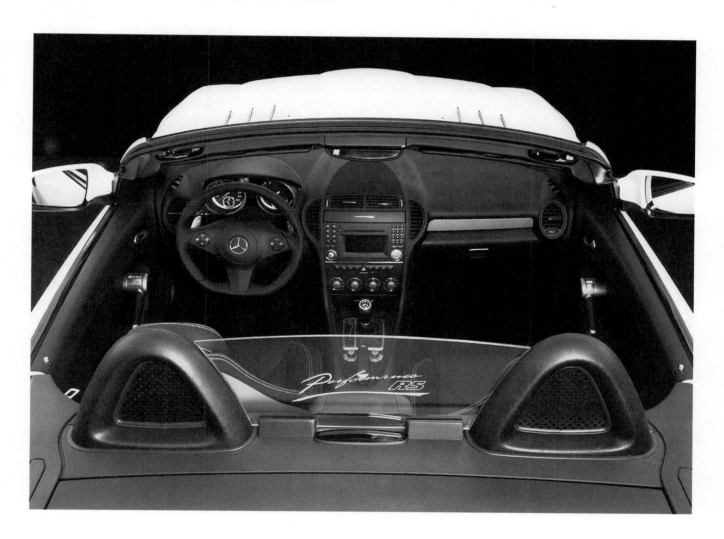

The best-known Mercedes tuner in the US is **RENNtech**, which was founded in 1989 by Hartmut Feyhl. Before forming RENNtech, Hartmut had worked with AMG, had become their technical director in the US and was responsible for the AMG Hammer, based on the W124 E-class cars. Today RENNtech does not concentrate on Mercedes cars only, but has specialized in several European automotive brands, among them Ferrari, Audi, Porsche and BMW.

RENNtech offered for the SLK various "Performance Packages". They started with the SLK230, where a larger crankshaft pulley for $1,195 was responsible for a power gain of 20 hp to 212 hp. Also the torque benefited with an increase of some ten percent to 300 NM (221 ft-lbs).

The most impressive performance hike was available for the SLK55 and called R2 package. For $15,900 it included RENNtech's proprietary ECU upgrade and a supercharger kit with intercooler upgrade, which changed the normally aspirated engine of the SLK55 to a supercharged version. Additional measures included a modified ignition timing and throttle mapping, an optimized air/fuel ratio and increasing the rev limiter. The result was dramatic to say the least, as it bumped up the output from 360 hp at 5,750 rpm to 460 hp at 5,500 rpm. Maximum torque was increased from 510 NM (376 ft-lbs) at 3,500 rpm to 630 NM (465 ft-lbs) available from 2,400 to 5,000 rpm.

The R2 package included a supercharger kit, which transformed the AMG SLK into a racing thoroughbred

For a mere $1,200, there was also a less dramatic option available, which was limited to tuning and upgrading the ECU. The result was an increase in output by 25 hp to 380 hp at 5,500 rpm. The peak torque was lifted by 40 NM (29.5 ft-lbs) to 535 NM (395 ft-lbs), which became available at a slightly lower 3,000 rpm. For $2,995, the R1 package increased power to 385 hp.

Similar packages were available for the SLK350. Another interesting option, especially for the SLK55, was for $3,690 the seven-speed transmission upgrade. It was a mechanical upgrade to the clutch packs with new gaskets, seals, o-rings and a new transmission pan filter and made gear changing even quicker.

The **C171 SLT** is probably a project that should not be taken too seriously, but Daimler-Benz watched of course the activities of BMW with its Z-series. When BMW had launched its Z3 Coupe, Daimler-Benz came up at the end of the 1990s with the design study of a potential R170 Coupe. Whether it was supposed to be called SLT is not known, but that was the name it was given by some German car magazines. In those days, it was even discussed that, should BMW offer an M-version of the Z3 Coupe, AMG would come out with its own SLT version.

This was one possible R170 SLT (sorry for the poor picture quality, the source was very small)

This was another option

At the end, those plans were dropped in favor of the C-class Sportcoupe, which became available in 2000. With the 2002 launch of the Z4, also a coupe was planned again and introduced in 2006:

Daimler-Benz could have possibly reacted with the following model. At least this was speculated around 2006 by the press:

149

The best advice is of course to have your used SLK inspected at an authorized dealer BEFORE any money changes hands. This will cost between $100 and $250 depending on the depth of inspection and will save you potentially not only a lot of frustration but more importantly also money. It is in general also advisable to buy a car with a properly documented service record.

The following list is by no means complete, but the author believes that it covers a reasonable share of issues that should be checked by a prospective buyer of the R171. Many of the points mentioned can be used for any car, but some are SLK specific. Quite a few of the potential problems can actually be fixed relatively easy, if one is a bit experienced. All points have been identified as either "Easy To Fix: ETF" or "Dealer To Fix: DTF". It needs to be mentioned here that the author takes no responsibility for any actions that result out of this. The recommendations given here should serve as a guide. That is why no prices for spares or repairs are given, as this varies from country to country. As already stated in an earlier chapter, please do not expect this to be a manual for car mainte-nance or repairs, because this is not this book's purpose. There are much better publications available. As the SLK shares many parts with the W203 C-class, Haynes publication in the UK can be a good source, if technical issues need to be addressed.

Body:

1. Although the second generation SLK is even better built than its predecessor and offers good resistance against corrosion, rust can be an issue in certain areas. Luckily most such occurrences are minor and non-structural. Rust should be checked on the front fenders, the rear wheel arches, the hood slam panel and the area under the trunk seals. Rarely corrosion can occur on the front corners of the hard top.

If you do find corrosion, give the car a very thorough inspection, especially of the underside. If such a car is at a location, where it cannot be lifted, do yourself a favor and walk away, unless you are willing to accept certain risks and can get the car at a bargain.

It is difficult to say, whether early R171s rust more often that later models, as it all depends on how the car has been treated. Naturally early R171s are older, so in theory, there should be more corrosion on them than on the later versions, but in general, early SLKs have the same rust prevention as newer models. ETF

2. Have a look inside the wheel-wells for loose or missing screws. ETF

3. Naturally you should have a look for any damage repair. One sign could be a poorly executed overspray. ETF

4. Open and close doors, trunk, hood and check for any leaks, worn out hinges, lack of support, etc. ETF

5. Are all badges/emblems still in place. ETF

6. If the car has xenon headlights, do they have both the same intensity? ETF

7. Check, whether trunk seals are good. If not, water can run into the pneumatic pump for central door locking. ETF

8. Ask, whether the car had an accident and if yes, was it repaired at an authorized service site such as the DB-dealer. Especially rear damage can be troublesome for the vario-roof, if not properly repaired. If in question, do walk away from that particular car, there are plenty others available. DTF

Not your average SLK. In Germany, companies like Carlsson offer such conversions

Wheels, brakes, exhaust:

1. Check tires for unusual or uneven wear. It can mean that front or rear wheels tracking can be off, which can cause, if neglected over a long period of time, expensive repairs to either front or rear suspension. It can also be a sign that the previous owner had pushed the car frequently pretty hard or used it often on drives over roads with pot holes. ETF

2. Another sign of a previously hard life for the car can be unusual wear of brake rotors and/or brake pad linings. DTF

3. Do alloys have scratched rim edges. If yes, it is mostly a cosmetic issue. Question is, can you live with this? ETF

4. In order to check for worn-out ball joints, grab the tires with both hands and push from side to side. They should not have any major movement. DTF

5. Is the spare still available with all its accessories such as the inflator. If there is no spare, is the TireFit kit still in the trunk. ETF

6. Drive the car on an empty road at some 10 to 15 km/h, then gently apply brakes with your hands lightly on the steering wheel. The SLK should hold its course without moving to one side. If not, the brake pads need replacement in most cases. ETF

7. While on the road, check the car's ESP system. Try to find a car park, where you can go for a spin. If the Brake Assyst (BAS) light comes on, there is a problem with the ESP. In most cases the electrical ESP motor of the system just needs a good cleaning. ETF

8. Drive the car on an empty road again and brake hard with both hands this time firmly on the steering wheel. If you can hear any knocking sound, the front suspension rubbers need replacement. This can be a common issue with cars over 80k to 100k km. DTF

9. Check the tailpipe for soot. If it's oily, it could mean that the rings are worn. DTF

Roof, trunk:

The roof subject has already been covered in the chapter that looks at roof problems, so here are just a few issues to be checked:

1. Most important, is the vario roof properly functioning. DTF

2. If yes, are the hinges and connections properly greased or lubricated. ETF

3. Hydraulic cylinders are expensive to replace, so check for any oil leaks, if possible also under the headliners (above the windscreen), as the front hydraulic cylinder tends to leak first. DTF

4. If the roof is not properly adjusted, it can lead to paint issues with the rear bumper and the trunk lid. DTF

5. Sometimes the glass windows in or behind one of the doors is not properly adjusted, which leads to water leaking between door glass and rear quarter window. DTF

6. This one is my favorite, although it is not directly related to the roof. I have a friend, who experienced some annoying rattling noise from the passenger side trunk area, every time he took the car out in winter or cooler days, when he had the roof closed. It did not occur with the top down. No MB dealer was able to fix it and it drove him nuts. Then I found a post on "slkworld.com", where a member had suffered from a similar issue with his car, but on the driver's side. So, all the credit for this solution goes with many thanks to "B55", a member of slkworld.com:

Just as I was about to give up I noticed something. The trunk rests on two pins at the inside upper right- and left-hand corners. These pins are long screws with flat round rubber heads. They are secured in place by nuts on the top and bottom of the steel plate that they are secured to. I noticed that the pin on the driver's side was significantly (about 3 inches!) higher than the passenger side pin. This caused the trunk top to close unevenly creating a much larger gap where the trunk seals when closed. I closed the trunk and pressed down on it hard a few times. I found that the gap on the driver's side had some play in it while the gap on the

passenger side did not. I concluded that the rattling was being caused by these pins being uneven. I was right!

I opened the top half way so that the nuts securing the pin were fully exposed. I used two sets of needle nosed pliers and loosened the nuts. I then lowered the driver's side pin so that it was exactly even with the passenger side pin. I lowered the driver's side pin because the passenger side wasn't making any noise. Once I tightened up the nuts again and closed the trunk, I found that there was no longer any disparity between the driver's side gap and the passenger side gap. Also, the play in the driver's side gap was gone. ETF

Engines, suspension:

1. The balance shaft problem has already been discussed extensively in an earlier chapter, so it does not need to be addressed here again.

2. Does the exhaust smoke blue, when the engine is running under load (this of course you can only realize, when you have a friend driving behind you). A dealer should check for cylinder leakage or determine, whether valve guides or rings are worn (ask for a compression test). Is there any burning smell under the hood?

This could also be caused by some oil runoff, or if the oil pan had been overfilled. Valves and rings are DTF

3. Does the exhaust smoke, when you drive the car and then slow down. It means that your rings could be worn. DTF

4. Start the cold engine and have someone close with his hand the exhaust tip. Have then someone check whether there is any leakage in the engine bay from the exhaust manifold. DTF

5. SLK200 Kompressor only: Does the engine run a bit rough? One suspect can be the MAF (Mass Air Flow) sensor. The MAF sensor samples incoming air after it has left the charged air cooler on its way to the throttle body. Depending where and how the car has been used, the MAF sensor's probe can become coated in dirt. ETF

6. Unfasten intake hose to the MAF. If you can see any signs of oil, the oil separator could have a problem. ETF

7. SLK200 Kompressor only: check for any noises from the supercharger. Any unusual whining could mean you need an expensive replacement. Do have this checked with a dealer before you buy that particular car. DTF

8. Any noise from the engine at cold start could come from one of two belt tensioner bearings. ETF

9. Check the drive belts for any wear. They should have been replaced on cars with over 100k km on the clock. ETF

10. When was the coolant changed? This should be done at least every three years. If you will do it, buy original coolant from your DB-dealer, it's not expensive. ETF

11. Depending on how hard the SLK was driven by its previous owner, at over 80K to 100k km the shock absorbers will be due for a change. So better check them. ETF

Transmission:

1. Automatic transmission and rear axle differential should be checked for oil leakage or noise. DTF

2. Ask the previous owner, when transmission fluid had been changed. It should be done every 30,000-40,000 km on a warm engine. ETF

3. If the transmission or differential is noisy, this is most often caused by worn or wrongly inflated tires. ETF

4. If you want to buy an SLK with manual transmission, this is how to check the clutch: after having started the car, put it in third gear and apply the brakes. Then slowly release the clutch, while still applying the brakes. If the clutch is ok, the car should stall now. If not, the clutch can be worn out and needs replacing. DTF

Check also the differential for any leakage

Interior:

1. Drive the car through a washing street and check for any leakage from the roof or windows. It could mean you have an issue with the hardtop or the windows, which do not fit precisely, as already mentioned earlier. DTF

3. In general the interior is fairly robust, so no major issues have been reported here, with the exception that some people have complained about the cup holders, whose location is directly above the center console in a superb place to spill coffee (or whatever the passengers like to drink) down the expensive COMAND electronics. Check also leather for wear. Most people will be unable to repair leather properly, so they are better advised to go to a specialist, so this is DTF

4. Do all the gauges function properly. DTF

5. When inside the car, turn all interior lights off. Switch on exterior lights and turn the dimmer switch to see, whether the instrument lights are working. ETF

6. Is the air-con working. DTF

7. Check if air circulation flap works by turning the fan on or off. It works vacuum-assisted and could be perforated. DTF

8. Do the electrical seats work in all directions, does the seat memory functions. DTF

9. Do heated seats work? If you knee on them in order to clean the car's interior, the heating system in the seat can break. The previous owner might have done so and it is expensive to repair. DTF

10. Do the glove box and center console lock and unlock with the doors

11. Does the COMAND work, is the screen free of defects and scratches

12. Are the nav data up-to-date (latest DVD)

13. Test the audio system with a CD

14. Is the steering wheel adjustment working

15. Is the AirScarf blowing warm air

16. Occasionally the AirScarf switches itself off after a few minutes. That could mean that the blower motor draws too much current and needs to be replaced

17. Very important: verify the VIN

The heating system in the seats can break and the AirScarf blower motor can draw too much current

Whoever designed this has never witnessed the detrimental effect hot coffee can have on expensive electronic devices

Make sure that all interior lighting incl. instruments work as advertised

The R171's VIN explained

The VIN (as part of the certification tag) is our SLK's unique DNA. All vehicles produced after January 1st 1989 have to have a "Vehicle Identification Number". In the US the VIN has been mandatory since the early 1980s already. It is always a unique 17-digit alpha and numeric string of characters for each car. The VIN is not something one should worry about, when buying a new car. But in case one plans to buy a second hand car it is best to spend a bit of money to use the VIN in order to get a vehicle history report. This report can inform the prospective owner, if the car in question had been previously stolen, whether there were any recalls, whether it had been wrecked, had any accidents, airbag deployments or mileage rollover to name just a few. It can even inform about the car's history, when it had been moved for example in the US from one state to another in order to wash the title clean.

In the case of Daimler-Benz, the VIN will start with "WDB"; for our SLK and cars of similar age. Sometimes also "WDC" was used (for Daimler-Chrysler). There are differences between the North American VIN's and the one for the rest of the world,

sometimes called FIN for "Fahrzeug-Identifikations-Nummer". For simplicity reasons, we will stick here to the name VIN.

The R171's VIN can be found near the lower left window post below the wiper. On US-bound cars the VIN can also be found on the builder's plate or as earlier said certification tag on the driver's door jamb above the lock (see photo on the next page). The certification tag covers next to the VIN also information about the car's color and weight:

The paint choice is a three-digit code to the right side of the VIN. It starts on our SLK and all other Mercedes cars of similar vintage with a letter before the code. Later cars had it higher up before the manufacturing date on the right side of the plate. Earlier Mercedes cars up to the 1970s had two paint suppliers, Glasurit and Herberts. In order to identify the respective supplier, either a "G" or "H" was placed in front of the paint code.

These characters are not used nowadays anymore. As can be seen from the certification tag on page 151, our sample SLK has a "C" in front of the paint code. It stands for a clear coat that was applied on top of all paints. Formerly also an "M" for metallic paint without clear coat or a "W" for water-based colors was used. At the end of 2003, Daimler-Benz started to use a new scratch-resistant clear coat lacquer made up of nano-sized ceramic particles, which are baked and hardened in the paint shop oven. It is a very hard paint finish, applied to all paints, not just metallic, that can better withstand mechanical car washes. In later years, the "C" was dropped.

The manufacturing date at the top right should be self-explanatory. The door jamb is also the place for the airbag replacement date. The engine number is stamped on the engine block behind the left cylinder head (driver's side).

On non-US-bound cars the VIN is stamped on the floor at the transmission tunnel behind the passenger seat behind a piece of carpet, which can be lifted. At the passenger's door jamb one can find the type plate and the paint code. On all cars the VIN can also be found in the inner wheel arch on the car's left side, the steering wheel column and of course in your SLK's title, guarantee and maintenance book.

159

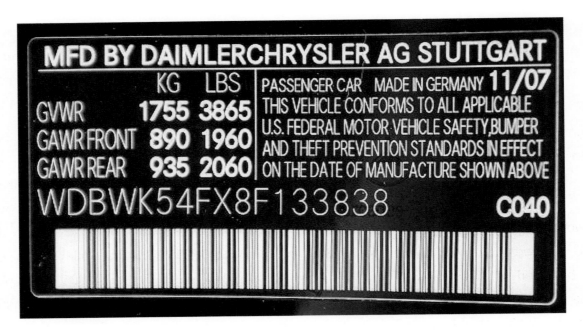

The certification tag (here for a US model, the European version will be explained afterwards) has next to the VIN its paint code, which is in our sample case "C040" for black.

Positions 1-3 identify the manufacturer. In our case it reads WDB, which stands for Daimler-Chrysler. The first character is used for the country of manufacture. "W" stands for Germany, "S" for the UK or "1" or "4" for the US

Position 4 identifies the respective Mercedes model and should read "W" for SLK, the SL R230 reads for example "S"

Position 5 identifies the body style. It reads "F" for a sedan version and "K" for the SLK and SL as Cabriolet/Roadster

Positions 6-7 identify the model within the series. "54" stands for the SLK280

Position 8 is meant for the safety restraint system, which in case of the SLK should read "F" as the car was equipped with side airbags

Position 9 determines through a some-what complicated mathematical formula that the previous numbers of the VIN are not fake

Position 10 represents the SLK's model year. It should not be confused with the year the car was actually delivered or sold. A "5" stands for 2005, while an "8" stands for 2008. At the end of the numerical counting (from 1 to 9) it starts with the alphabet, so "A" stands for the year 2010 for example.

Position 11 identifies the plant, where the car was manufactured. In our case it should be a letter from "F-H", which indicates for both the SLK and SL the plant in Bremen. Letters A to E are used for the plant in Stuttgart-Sindelfingen.

Positions 12-17 indicate the order in which the SLK has left the assembly line. This is independent of the car's engine. The last four of these characters are always numeric.

These last six digits are also the SLK's chassis number. They are perhaps the most critical numbers of the VIN. Due to possible mid-year production changes, they are vital in identifying the proper part numbers for ignition, fuel, emission and engine components. Such parts are often listed with the caveat that they fit SLK models up to a particular VIN or before or after a VIN sequence.

If we use this particular US-bound VIN, the respective FIN for that car would look as follows:
WDBWK54FX8F133838
changes into:
WDB171454FX8F133838
As one can see, FINs for cars outside the US start after the country and manufac-

turer's code with the type of car (171), followed by the version, which is in our case for the SLK280 "454". More information on the chassis prefixes one can find on the next page. The remaining digits have the same meaning as on the US-bound vehicles.

The certification tag of the R171 does not contain any information regarding the interior or features/extras of the car. There are various sites on the internet, where, after insertion of the VIN, exact data on a particular car's equipment will be given.

Information about the paint and a car's specific equipment can usually be found on the data card that came with the car, when it was new. On earlier Mercedes models a plate with respective paint and body information was located prominently on the right side of the radiator cross support. As we have seen, with the R171 and other newer Mercedes models, this is now different.

This chapter is a good place to list the respective internal code numbers of all R171 versions:

1.*SLK200 Kompressor R171 E18*, built from March 2004 to March 2008:

Chassis prefix: 171.442, engine M271 E18 prefix: 271.944

2.*SLK350 R171 E35*, built from March 2004 to March 2008:

Chassis prefix: 171.456, engine M272 E35 prefix: 272.963

3.*SLK55 AMG R171 E55*, built from Sept. 2004 to March 2011:

Chassis prefix: 171.473, engine M113 E55 prefix: 113.989

4.*SLK280/300 R171 E30*, built from April 2005 to March 2011:

Chassis prefix: 171.454, engine M272 E30 prefix: 272.942

5.*SLK200 Kompressor R171 E18*, built from April 2008 to March 2011:

Chassis prefix: 171.445, engine M271 E18 prefix: 271.954

6.*SLK350 R171 E35*, built from April 2008 to March 2011:

Chassis prefix: 171.458, engine M272 E35 prefix: 272.969

The R171's data card

The Owner's Manual and Service Booklet should ideally still be with the car, as they contain two data cards. One of these cards contains the key codes and should definitely NOT be stored in your car. The second one does not contain this information and should be kept in the car, as it helps the technician in the service facility to order the correct parts. The data card has codes for the original equipment and additional extras that came with the car.

Everything printed on it is in German, so on another page it offers some translations and space for the owner to write down his name, address etc. If the car`s Data Card is lost, one of the DB- or SLK-enthusiast websites could be of assistance. Search those sites for key-words such as "Factory Option Codes" or "Data Card". That will lead you to fellow enthusiasts, who might be able to help retrieving the Data Codes from the VIN of your car. Another option to retrieve those data can be a drive to your DB-dealer or Mercedes specialist shop. They can use their diagnostic scan tool to get the info from the vehicle's ECU (Engine Control Unit).

Knowing the options of your car, when it left the assembly line, might be quite helpful, when ordering parts for it. Sometimes you will find that the optional extras, your SLK should have come with, are not anymore there. One example is the CD changer. Even when there is none available, a quick look into the glove compartment might reveal its harnesses, so a replacement could be fairly simple.

The upper half of the data card carries 40 number spaces (as one can see on the next page, not all of them are filled out). They show from the chassis-, order-, production-, transmission- numbers etc. all basic information that is important to your specific vehicle.

The three-digit numbers that can be found below them are of equal importance as they define the options that came with your specific SLK (and any other Mercedes as a matter of fact), when it left the assembly line. It was easy for North America, as the SLK came almost fully equipped with just five options to choose from.

But the situation was very much different for the rest of the world, where the car came in many countries standard with fabric seats, no air-con or adjustable steering column to name just three.

As the codes on the data card might be a bit difficult to read in a book, we will look at it box by box. In case you find this subject a bit too boring, please skip it and go to the next chapter that deals with the car's second-hand prices. We start with the upper part of the data card, left side: Our sample data card is from an SL R230, which was built around the same time, the R171 has been produced. Again, this data card is not from an actual SL and should only serve as reference

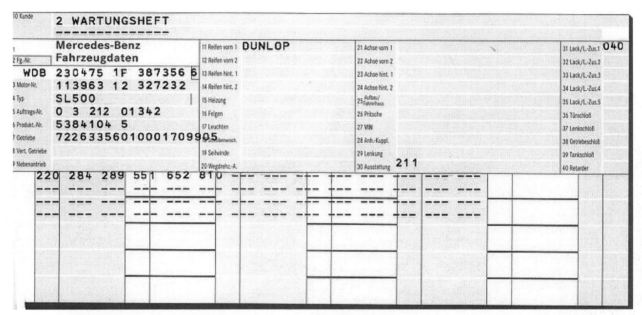

A sample data card from an SL500 from 2003

Box 2, which contains the VIN, does not need to be explained again.

Box 3 shows the engine number, which follows more or less the same logic as the chassis no or VIN. It starts with 113 for the type of engine, which is followed by the specific engine version.

The M113 came in various versions, from 4.3 L to 5.5 L and was used in many Mercedes models from the CLK430 to the R500. It was even the basis for the DTM CLK racecar and as already mentioned earlier the SLR McLaren.

The seventh digit identifies "1" lhd or "2" rhd cars, while the eighth digit shows the engine mounted to a car with "0" manual or "2" automatic transmission. In case you wonder, why "0" and not "1" was used for the manual car, the "1" had been used in the late 1950s for the "Hydrak-system". The last six digits show that in our case the engine was the 327,232nd one produced. They are sequential by transmission type, so counted separately for manual and automatic versions in case of cars that offered such a choice.

Box 5 shows the "Auftrags-Nr" or order no. It is one of the most interesting codes of the data card and consists of ten digits. The second digit indicates the last digit of the year the car was ordered, in our case 2003. The next three digits show, which dealer/country has placed the order for that car. If the code starts with a "2" or "3", it shows that the car was ordered by a Mercedes subsidiary (or its affiliate) in Germany. Then the following two digits determine together with the "2" the region and dealer. Our SL shows the region code "212", which meant that it was ordered by the Mercedes subsidiary in Berlin.

In case of places outside Germany, a "5" as first digit means countries in other parts of Europe. Code "537" means the UK, "531" France, "577" Spain or "543" Italy for example.

Countries in the Americas start with a "7". If followed by "03" to "07" they were destined for the US (for a short time until 1965 also codes 708 to 718 were used for the US). As a side note, an interesting number is "707" as it means that this particular car was bought through a so-called European Delivery Program, which was quite popular in the 1960s and 1970s. Canada uses code "701". The Austral-Pacific region starts with a "9" (901 for Australia, 919 for NZ), while markets in Asia have an "8" as first of these three digits (Japan: 839, Hong Kong: 823).

The remaining four digits are sequential for that particular region/dealer and that year.

Box 6 indicates the "Prod.-Nr." or "Produktions-Nummer", which is the car's production number

Box 7 shows the transmission number, which is also stamped inside the bell-housing.

The upper middle and right side of the data card (please see picture below) offer in comparison usually relatively little information, although there is in theory plenty of space for all kind of equipment. Alone the first four boxes 11 to 14 can be used for tire-related information, which shows that cards like this are being used for all kind of Mercedes vehicles, from the Unimog to the largest truck. The remaining boxes 15 to 20 can provide info regarding heating, rims, lights (usually Bosch or Hella), wipers down to winches and speedometers.

The upper center part starts with four boxes reserved for front and rear axles.

Box 25 is the "Aufbau-Nr." or body number and not filled-out in our example. It was necessary during the production process to link all the parts before they finally became a complete car and before a "VIN" was assigned to the vehicle. One can see this number also on various other parts such as hood, trunk lid or inner side of doors panels for example. This code was always part of the data card on older cars, now it is used primarily for trucks. That is why the name "Fahrerhaus" (driver cabin) has been added.

11 Reifen vorn 1 **DUNLOP**	21 Achse vorn 1		31 Lack/L.-Zus.1 **040**
12 Reifen vorn 2	22 Achse vorn 2		32 Lack/L.-Zus.2
13 Reifen hint. 1	23 Achse hint. 1		33 Lack/L.-Zus.3
14 Reifen hint. 2	24 Achse hint. 2		34 Lack/L.-Zus.4
15 Heizung	25 Aufbau/Fahrerhaus		35 Lack/L.-Zus.5
16 Felgen	26 Pritsche		36 Türschloß
17 Leuchten	27 VIN		37 Lenkschloß
18 Scheibenwisch. **05**	28 Anh.-Kuppl.		38 Getriebeschloß
19 Seilwinde	29 Lenkung		39 Tankschloß
20 Wegdrehz.-A.	30 Ausstattung **211**		40 Retarder

The same applies to box 26, where "Pritsche" is the German word for flat-bed. The next three boxes 27 to 29 cover the VIN, towbar (Anhängerkupplung) and steering (Lenkung). The steering – box was also always filled out on older data cards, now it does not seem to be needed anymore, at least not for passenger cars.

Box 30 indicates the "Ausstattung" or interior choice. If that number starts with a "2" like in our case, the interior comes in leather, a "1" would stand for an MB-Tex interior, while a "0" means fabric upholstery, both not available in our SL case.

Box 31 at the top right side shows the paint code. There is no code for the paint of the plastic parts of the lower body.

The lower portion of the data card shows all options that were ordered for the car, when new. Most of these codes have seen over the years numerous assignments.

In case of modern Mercedes cars, the first code "220" to the left stands for the Parktronic feature. Earlier it was used for rear door contacts. Also code "284" is a good example. Now it shows the gear lever in wood/leather combination. In the 1960s and 1970s it was used for a five-piece luggage set. A more drastic example would be code "317", which was used for the Comfort Package. Quite a few years back that code was used for cars that were sent to Portugal as driver's school vehicles and thus as used cars most probably best to be avoided. ☺

In rare cases, these codes have seen new assignments even during production of a certain model, as has happened with the R107.

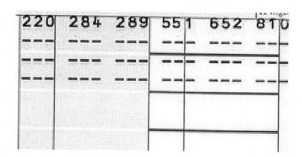

167

The R171's option codes

The following code list might be a bit long and if you think I should take it out again, I would appreciate your feedback. On the other hand some of you might find it quite handy, as it covers all options that were available for the R171. With regards to country codes, I have only listed the codes for the major countries, in order not to make this list too long. I don't think that anybody here would need the code for Morocco or Guinea-Bissau. The code system comes in three parts. The first one has only numbers and deals mostly with technical equipment, the second part has numbers, followed by a letter and covers most of the paints. The third part starts with a letter, followed by numbers, it covers equipment parts such as designo, special wheels etc. But don't be surprised to find also technical extras under the last part or designo extras in the first part. The three parts are separated by pictures.

171 003 MANUAL DATE-CONTROLLED VEHICLE

171 004 INSPECTION VEHICLE

171 005 VEH. FOR CERTIFICATION/HOMOLOGATION/TYPE APPROVAL

171 006 SPECIAL TEST VEHICLE

171 007 TEST DEPT. SINDELFINGEN, BODYWORK DEVELOPMENT

171 008 NEAR-LAUNCH ROAD TRIALS, FLEET UNTERTUERKHEIM /TRI

171 009 PHOTO

171 010 ROAD TEST W50

171 011 ROAD TEST W67

171 012 TRAINING / DIAGNOSIS VP/S

171 013 EXHIBITION (IF NOT CODE 997)

171 014 NEAR-LAUNCH ROAD TRIALS, FLEET SINDELFINGEN/W54 RA

171 015 KD - EQUIPMENT

171 016 FIELD TEST INLAND/EXPORT

171 017 NON CUSTOMER READY VEHICLE

171 018 CUSTOMER READY VEHICLES WHICH SHOULD BE MODIFIED

171 019 FIELD TEST VEHICLES, DEMONSTRATION VEHICLES (CUSTO

171 020 DEMONSTRATION VEHICLE, SALES

171 021 DESIGNO - LEATHER

171 022 DESIGNO - DECORATIVE TRIM

171 024 QM RELEASE VEHICLES

171 027 VEH. W/O CAMOUFLAGE (CONTRARY TO CAMO. REGULATION)

171 030 ELECTR. REGULATED DIESEL FUEL PUMP

171 032 TAIWAN - GERAEUSCHTEST GETRIEBE

171 033 PORTABLE CTEL CONVERSION (FROM NOKIA TO SIEMENS)

171 070 MAJOR ASSEMBLY FOR TEST VEHICLE

171 071 MAJOR ASSEMBLIES FOR MEASUREMENT VEHICLE

171 100 CONTROL CODE NULLSERIES VEHICLE

171 101 CONTROL PRODUCTION TEST 1

171 102 CONTROL PRODUCTION TEST 2

171 103 CONTROL PRODUCTION TEST 3

171 104 SERVICE VEHICLES AT STARTUP

171 105 VEHICLES FOR RALLY

171 106 CONTROL CODE MARKETING

171 107 CONTROL CODE MARKETING

171 108 CONTROL CODE MARKETING

171 109 CONTROL CODE MARKETING

171 111 STEUERCODEREIFEGRADABSICHERUNG 1 RGA1/NULLSERIE

171 112 CONTROL CODE ENSURE OF MATURITY 2 RGA2

171 113 CONTROL CODE ENSURE OF MATURITY 3 RGA3

171 114 CONTROL CODE CUSTUM-PRODUCT-AUDIT CPA

171 116 STEUERCODE REIFEGRADABSICHERUNG 4 RGA4

171 140 PACKAGE 1

171 141 PACKAGE 2

171 142 PACKAGE 3

171 143 PACKAGE 4

171 170 ENGINE OUTPUT, FRONT

171 202 OWNERS MANUAL AND SERVICE RECORD - GERMAN

171 205 OWNERS MANUAL AND SERVICE RECORD - ENGLISH

171 206 OWNERS MANUAL AND SERVICE RECORD - ITALIAN

171 207 OWNERS MANUAL AND SERVICE RECORD - FRENCH

171 208 OWNER'S MANUAL AND SERVICE BOOKLET - IN SPANISH

171 209 OWNER'S MANUAL AND SERVICE BOOKLET - IN PORTUGUESE

171 213 SPEED-SENSITIVE POWER STEERING/VARIO STEERING

171 219 PROXIMITY-CONTROLLED CRUISE CONTROL

171 220 PARKTRONIC SYSTEM (PTS)

171 221 LEFT FRONT SEAT, ELECTRICALLY ADJUSTABLE

171 222 RIGHT FRONT SEAT, ELECTRICALLY ADJUSTABLE

171 231 GARAGE DOOR OPENER

171 232 GARAGE DOOR OPENER WITH 284 - 390 MHZ FREQUENCY

171 236 DAYDRIVINGLIGHT

171 241 FRONT SEAT LH ELECTRIC ADJUSTABLE WITH MEMORY

171 242 FRONT SEAT RH ELECTRIC ADJUSTABLE WITH MEMORY

171 249 INSIDE AND OUTSIDE MIRROR AUTOMATIC DIMMING

171 250 AMG DRIVER PACKAGE

171 251 INCREASED SPEED LIMIT FOR NAFTA

171 254 RADIO MB AUDIO 30 - USA

171 255 RADIO MB AUDIO 30 - JAPAN

171 260 TYPE DESTINATION ON TRUNK LID – ELIMINATION

171 261 TYPE DESIGNATION ON FENDER / SIDEWARD DELETION

171 263 LICENSE PLATE ATTACHMENT ASIA / MEXICO

171 264 LICENSE PLATE ATTACHMENT AMERICA

171 267 TYPE-DESIGNATION DEVIATING

171 270 INVALID/ANTENNA F.D-NET TELEPHONE ON FENDER REAR.L

171 273 TELEPHONE PRE-INSTALLATION D-NET MOBILE

171 275 MEMORY PACKAGE (DRIVER SEAT, STRG. COL., MIRROR)

171 278 TELEPHONE PRE-INSTALLATION COMPL. D-NET

171 280 LEATHER STEERING WHEEL AND LEATHER GEAR SHIFT KNOB

171 281 AMG PERFORMANCE STEERING WHEEL

171 283 DRAFT STOP/GLASS DRAFT STOP

171 285 DRAFT STOP (FABRIC DESIGN)

171 289 WOOD/LEATHERSTEERING-WHEEL

171 294 KNEEAIRBAG

171 301 ASHTRAY PACKAGE

171 309 CUP HOLDER

171 312 TELEPHONE D-NET "MOBILE" AT TOWER (NOKIA 3110)

171 315 PHONECARD D1 DEBITEL (PRICE REDUCTION)

171 316 INVALID/ TELEPHONE (D2B) CENTER CONSOLE (MOTOROLA)

171 317 TELEPHONE (D2B) HANDY IN CENTRAL CONSOLE

171 318 CELL PHONE IN GLOVE BOX

171 328 PHONECARD D2 DEBITEL (PRICE REDUCTION)

171 329 TELEPHONE PRE-INSTALLATION AMPS

171 330 CD COMPARTMENT

171 331 CASSETTE COMPARTMENT

171 343 AIR FILTER

171 345 RAIN SENSOR

171 347 AUTOMOBILE TELEPHONE "TELE AID" FITTED SYSTEM

171 348 TELEAID LOW COST USA INCL.MOBILE PRE-INSTALLATION

171 349 PREP. FOR EMERGENCY CALL SYSTEM

171 352 COMAND

171 353 AUDIO 30 APS

171 358 TELEAID LOW COST - ECE

171 359 LOW COST TELEAID USA

171 380 COMPLETE CELL PHONE PREINSTALLATION

171 381 CTEL FIXED INSTALLATION CP LIGHT

171 382 PORTABLE CTEL

171 383 CTEL FIXED INSTALLATION CP HIGH

171 384 TELEPHONE PREINSTALLATION CP HIGH FOR USA

171 385 TELEPHONE PREINSTALLATION CANADA/AMPS

171 386 MOBILE PRE.-INSTALLATION (MTUS)

171 387 MOBILE PRE.-INSTALLATION (MTUS) USA / CANADA

171 388 PHONE MOBILE, MTUS-SYSTEM

171 389 MOBILE PRE.-INSTALLATION (MTUS) USA/CANADA COMPL.

171 390 LINGUATRONIC-ENGLICH(BRITISH)

171 391 LINGUATRONIC-FRENCH

171 392 LINGUATRONIC-SWISS GERMAN

171 393 LINGUATRONIC-ITALIAN

171 394 LINGUATRONIC-SPANISH

171 401 FRONT SEAT CLIMATE CONTROL

171 403 AIRSCARF SYSTEM

171 411 6-SPEED MANUAL TRANS.

171 423 5-SPEED AUTOMATIC TRANSMISSION

171 424 SEQUENTRONIC

171 426 CVT - AUTOMATIC TRANSMISSION

171 427 AUTOMATIC TRANSMISSION 7-SPEED

171 428 STEERING WHEEL SHIFT BUTTONS/GEARSHIFT PADDLES

171 440 TEMPOMAT (CRUISE CONTROL)

171 441 STEERING COLUMN, ADJUSTABLE

171 443 HEATED STEERING WHEEL

171 460 KANADA-AUSFUEHRUNG / ZUSATZTEILE

171 461 INSTRUMENT WITH MILES IND. AND ENGLISH LEGEND

171 470 TIRE PRESSURE MONITOR (TPM) LOW LINE

171 475 TIRE PRESSURE MONITOR (TPM) HIGH LINE/MID LINE

171 477 TIRE PRESSURE LOSS WARNER

171 481 UNDERSHIELDS

171 485 (COMFORT RUNNING GEAR)

171 486 SPORTS SUSPENSION

171 494 U.S. VERSION

171 496 MEXICO VERSION

171 498 JAPAN VERSION

171 500 OUTSIDE REAR VIEW MIRROR LH AND RH FOLDING

171 510 AUDIO 20 RADIO WITH CD CHANGER

171 511 AUDIO 50 APS RADIO WITH DVD CHANGER

171 512 COMAND APS WITH DVD CHANGER

171 517 HDRADIO(HIGHDEFINITION)

171 518 UNIVERSAL COMMUNICATIONS INTERFACE (UCI)

171 523 RADIO AUDIO 20

171 524 PRODUCT PROT. F.PAINT - TRANSPORT VEH.PRESERVATION

171 525 RADIO AUDIO 50 APS

171 526 COMAND DVD (WITHOUT NAVIGATION)

171 527 COMAND DVD APS WITH NAVIGATION

171 528 COMAND DVD PLUS (WITHOUT NAVIGATION)

171 529 COMAND DVD JAPAN WITH NAVIGATION

171 530 COMAND DVD APS USA WITH NAVIGATION

171 531 SIRIUS SATELLITE RADIO FOR USA PREINSTALLATION

171 533 SPEAKERS FRONT AND REAR - WITHOUT RADIO

171 534 AUDIO 20 CD RADIO, JAPAN

171 535 AUDIO 20 CD RADIO, USA

171 536 SIRIUS SATELLITE RADIO COMPLETE SYSTEM

171 537 DAB TUNER

171 543 SUN VISOR WITH MAKE-UP MIRROR L. AND R., BE-

171 551 ANTI-THEFT/ANTI-BREAK-IN WARNING SYSTEM

171 560 ELECTRIC. ADJUSTABLE DRIVER SEAT L. AND R.

171 580 AIRCONDITIONER

171 581 AUTOMATIC CLIMATE CONTROL

171 596 HEAT INSULATING+IR REFLECTG.SAFETY GLASS ALL-ROUND

171 600 HEADLAMPS - CLEANING EQUIPMENT

171 602 FRONT MOUNTED LICENSE PLATE BRACKET

171 606 L.A.WHL 5-HOLE DESIGN 16 INCH WITH MIXED TIRES

171 607 L.A.WHL STAR-DESIGN 17 INCH WITH MIXED TIRES

171 611 EXIT LIGHTS FOR DRIVER DOORS

171 612 HEADLAMP-XENON R.H.TRAFFIC

171 613 HEADLAMP LEFT-HAND TRAFFIC

171 614 BI-XENON HEADLAMPS FOR RIGHT-HAND TRAFFIC

171 617 XENON HEADLAMPS, LEFT-HAND TRAFFIC

171 618 BI-XENON HEADLAMPS FOR LEFT-HAND TRAFFIC

171 619 CORNERINGILLUMINATION

171 623 VERSION FOR THE GULF STATES

171 625 VERSION FOR AUSTRALIA

171 634 DELETION - FIRST AID KIT

171 636 DELETION - WARNING TRIANGLE

171 639 LA 5 SPOKE DESIGN 17" WITH MIXED TIRES

171 645 WINTER TIRES M + S

171 651 MOUNTED WINTER WHEELS

171 652 SUMMER TIRES MOUNTED

171 656 LIGHT ALLOY WHEELS, 10-HOLE

171 657 L.A.WHLS, 8-HOLE WITH WIDE TIRES

171 666 PRODUCT PROT.F.TRANSPORT VEH. W/O TIE-DOWN HOOKS

171 668 PRODUCT PROT.F.TRANSPORT VEH. W/ TIE-DOWN HOOKS

171 669 SPARE WHEEL / FOLDING WHEEL

171 670 RESIDUAL ENGINE HEAT UTILIZATION (MRA)

171 673 HIGH-CAPACITY BATTERY

171 682 FIREEXTINGUISHER

171 685 LIGHT ALLOY, 7-HOLE DESIGN, 16 INCH

171 687 MODEL YEAR ON TYPE LABEL

171 688 INVALID/L.A.5-SPOKE DESIGN,16" WITH MIXED TIRES

171 721 BASIC CARRIER

171 728 VAVONA WOOD FINISH, INC. STEERING WHEEL

171 729 POPLAR WOOD VERSION

171 734 EUCALYPTUS WOOD FINISH

171 736 WOOD DESIGN ASH-TREE BLACK

171 751 AMG FACTORY DELIVERY

171 753 RADIO MB AUDIO 10 CC WITH VK/RDS

171 756 RADIO MB AUDIO 10 CD WITH VK/RDS

171 761 RADIO RC W. REDUCED RANGE W/O PANIC (315 MHZ)

171 762 RADIO REMOTE CONTROL W/O PANIC SWITCH (315 MHZ)

171 763 RADIO REMOTE CONTROL W/ PANIC SWITCH (315 MHZ)

171 767 AMG DISK WHEEL 17 INCH WITH MIXED TIRES

171 772 AMG STYLING PACKAGE-FRONT SPOILER, SIDE SKIRT

171 779 AMG DOUBLE-SPOKE WHEELS 17" WITH MIXED TIRES

171 782 AMG-SPOKEN WEELS 18" MIXED TIRES

171 784 AMG 17" SPOKE WHEEL MIXED TYRES

171 786 AMG DOUBLE-SPOKE WHEELS 18" WITH MIXED TIRES

171 794 AMG-WHEELS 7,5 X 17 SQUEEZE-IN DEPTH 37 VA 225/45

171 795 AMG 18" DOUBLE-SPOKE WHEELS WITH MIXED TIRES

171 796 AMG 18" WHEELS, MULTISPOKE DESIGN W. MIXED TIRES

171 810 SOUND SYSTEM

171 813 LINGUATRONIC-GERMAN (VOICE OPERATION)

171 817 HOUSING FOR ALARM SIREN (ATA)

171 819 CD CHANGER

171 820 VEHICLES FOR TOURIST

171 822 VEHICLES FOR SOUTH-AFRIKA, ADDITIONAL PARTS

171 823 VEHICLES FOR SWITZERLAND, ADDITIONAL PARTS

171 824 ADDITIONAL PARTS FOR COUNTRIES WITH COLD CLIMATES

171 825 VEHICLES FOR SWEDEN, ADDITIONAL PARTS

171 829 VEHICLES FOR NORWAY, ADDITIONAL PARTS

171 830 CHINA-VEHICLES-ADDITIONAL PARTS

171 831 VERSION FOR ITALY SUPPLEMENTARY PARTS

171 832 VEHICLES FOR DENMARK, ADDITIONAL PARTS

171 833 INVALID/GREATBRITAIN, ADDITIONALPARTS

171 835 SOUTHERN KOREA-SUPPLEMENTARY PARTS

171 836 ADDITIONAL PARTS FOR SINGAPORE

171 852 MOBILE.PRE-INSTALL.COMPL.ON TOWER NOKIA SERIES 5/6

171 853 TELEPHONE FITTED ON "TOWER"; NOKIA, D-NET

171 854 TELEPHONE "CELL PHONE", NOKIA 6210/6310I

171 855 TEL. TELEAID AT "UPPER CENTER CONSOLE",NOKIA,D-NET

171 859 CTEL CARD E-NETWORK DEBITEL (PRICE REDUCTION)

171 862 PROVISION FOR INSTALLATION OF TELEVISION

171 865 DIGITAL TV FOR JAPAN

171 873 SEAT HEATER FOR LEFT AND RIGHT FRONT SEATS

171 875 HEATED SCREEN WASH SYSTEM

171 876 INTERIOR LIGHT ASSEMBLY

171 880 CLOSING SYSTEM WITH INFRARED REMOTE CONTROL

171 882 INTERIOR SAFEGUARD

171 885 HIGHTEN THEFT PROTECTION

171 887 SEPARATE TRUNK LID LOCKING

171 889 KEYLESS - GO

171 911 MAJOR ASSEMBLIES FOR SPECIAL TEST VEHICLES

171 918 EMISSION CONTROL SYSTEM UNREGULATED

171 922 POWER REDUCTION, ENGINES

171 928 EXHAUST GAS CLEANING WITH EURO 5 TECHNOLOGY

171 951 SPORT-PACKAGE USA

171 952 SPORTS PACKAGE

171 966 COC PAPER EURO 5 TECH. W/O REGIST. CERT. PART 2

171 967 COC PAPER EURO 5 TECH. W. REGISTRAT .CERT. PART 2

171 980 REG. DOC. AND ABE NO. ON TYPE LABEL FOR EXPORT

171 981 TUEV INDIV. ACCEPTANCE, DOMESTIC, W/O COC PAPERS

171 982 VEHICLES FOR FINLAND, ADDITIONAL PARTS

171 983 COC PAPER EURO 4 TECHNOLOGY W/O REG. CERT. PART II

171 984 COC DOC., DELETION OF VEHICLE REGISTRATION DOC.

171 985 COC PAPER EURO 4 TECHNOLOGY W. REG. CERT. PART II

171 986 IDENTIFICATIONNUMBER (VIN-NO.)

171 987 BATTERY ISOLATION SWITCH FOR SHIPMENT VEHICLES

171 988 VEH. TIT. OF OWN. & COC DOC. + MODEL PLATE (MACH.)

171 989 IDENTIFICATION LABEL UNDER WINDSHIELD

171 990 AMG - VEHICLE DOCUMENTS/MODEL PLATE MANAGEMENT

171 994 STATIONARY CAR FOR USE ON EXHIBITION

171 997 STATIONARY CAR FOR USE ON EXHIBITION

171 000A FABRIC

171 011A FABRIC - BLACK/ANTHRACITE

171 016K ROTARY VALVE THERMOSTAT

171 01P CAMPAIGN VEHICLES, DCVD

171 01Q GROSSVERSUCHKEILRIPPENRIEMEN

171 021A FABRIC - BLACK/ANTHRACITE

171 022A FABRIC - BLUE

171 025U DESIGNO-BRILLIANT BLACK PAINT

171 026A FABRIC - GREEN

171 027A FABRIC - RED

171 029U DESIGNO-SILVER PAINTWORK

171 02P VEHICLE NOT FOR PLANNING

171 02T MAJOR ASSEMBLY FOR ADVANCED PROTOTYPE

171 031U DESIGNO-CHROMAFLAIR 1 PAINTWORK

171 032U DESIGNO-MYSTIC CLUE PAINTWORK

171 033U DESIGNO-MOCCA BLACK PAINTWORK

171 034U DESIGNO VARICOLOR 3 PAINT COATS

171 036U DESIGNO VARICOLOR 4 PAINT COATS

171 037U DESIGNO MYSTIC RED PAINT

171 038U DESIGNO MYSTIC WHITE PAINT

171 039U DESIGNO MAURITIUS BLUE PAINT

171 03P VEHICLE NOT FOR PLANNING

171 03T STEUERCODE FUER SPEZIFISCHE ENTWICKLUNGSFAHRZEUGE

171 040U BLACK

171 041U DESIGNOGRAPHITEPAINTWORK

171 042U DESIGNO-HAVANNA PAINTWORK

171 043U DESIGNO-CHABLIS PAINTWORK

171 044U MAGNO ALLANITE GRAY PAINT

171 045U DESIGNO SABBIA MAGNO PAINTWORK

171 046U DESIGNO-PLATINUM BLACK PAINTWORK

171 048U DESIGNO MYSTIC WHITE 2 PAINT

171 049U DESIGNO MAGNO KASHMIR WHITE PAINTWORK

171 04P VEHICLE NOT FOR PLANNING

171 051U DESIGNO MAGNO PLATINUM MATT FINISH

171 052U DESIGNO MYSTIC BROWN PAINT

171 05P VEHICLE NOT FOR PLANNING

171 06P MILEAGE

171 07Q GV-NAOBRAKE LININGS

171 080A FABRIC

171 081A FABRIC - BLACK/ANTHRACITE

171 08P CONTROL CODE MARKETING

171 09P CONTROL CODE MARKETING

171 10P CONTROL CODE MARKETING

171 13P FAIR PLAY MODEL

171 14P ME VEHICLE BEFORE DELIVERY CONFIRMATION

171 15P ME VEHICLE AFTER DELIVERY CONFIRMATION

171 16P ME VEHICLE

171 189U GREEN BLACK METALLIC

171 197U OBSIDIAN BLACK

171 1R4 COMPLETE WINTER WHEEL, L. ALLOY 5-SPOKE DESIGN 16"

171 1R5 COMPLETE WINTER WHEEL, L. ALLOY 5-SPOKE DESIGN 17"

171 1R6 COMPLETE WINTER WHEEL, L. ALLOY 7-SPOKE DESIGN 16"

171 1R7 COM. WINT. WHEEL, L. ALLOY 5-DOUBLE-SPOKE DES. 17"

171 1U3 VEHICLE WAXING

171 1U4 VEHICLE WAXING WITH ADDITIONAL SCOPE

171 200A LEATHER

171 201B OWNERS MANUAL AND SERVICE RECORD - DUTCH

171 202B OWNERS MANUAL AND SERVICE RECORD - GERMAN

171 204B OWNERS MANUAL AND SERVICE RECORD - FINNISH

171 205B OWNERS MANUAL AND SERVICE RECORD - ENGLISH

171 205K NWG-WANDLER 0 MM B.ST.0

171 206B OWNERS MANUAL AND SERVICE RECORD - ITALIAN

171 207B OWNERS MANUAL AND SERVICE RECORD - FRENCH

171 208B OWNERS MANUAL AND SERVICE RECORD - SPANISH

171 209B OWNERS MANUAL AND SERVICE RECORD - PORTUGESE

171 20P NAVIGATION COMFORT PACKAGE

171 210B OWNERS MANUAL AND SERVICE RECORD - SWEDISH

171 211A LEATHER - BLACK/ANTHRACITE

171 211B OWNERS MANUAL AND SERVICE RECORD - ARABIC

171 212B OP. INSTRUCT. AND MAINT. BOOKLET-ENGL. FOR USA/CAN

171 220B INFO PLATE, COOLANT - GERMAN

171 221A LEATHER - BLACK/ANTHRACITE

171 221B SIGN COOLANT/REFUELING - ENGLISH

171 222A LEATHER - BLUE

171 222B INFO PLATE COOLANT/REFUELING - FRENCH

171 223B INFO PLATE, COOLANT - SPANISH

171 224B COOLANT WARNING SIGN IN PORTUGUESE

171 225A LEATHER - BEIGE

171 227A LEATHER - RED

171 229L GERMANY

171 22R 18" LIGHT ALLOY 5 DBL.-SPOKE DESIGN W. MIXED TIRES

171 230B CUSTOMER SERVICE - STATION EUROPE

171 230L GERMANY - IMPORT

171 230U INDIGOLITEBLUEMETALLICFINISH

171 231B CUSTOMER SERVICE - STATION LATIN AMERICA

171 232B CUSTOMER SERVICE - STATION ASIA WITH ISRAEL

171 233B CUSTOMER SERVICE - STATION ASIA WITHOUT ISRAEL

171 234B CUSTOMER SERVICE - STATION AFRICA

171 235B CUSTOMER SERVICE - STATION AUSTRALIA

171 236B WITHOUT CUSTOMER SERVICE - STATION INDEX

171 250B MOBILO LIFE/EUROPE SERVICE PACKAGE

171 251B DIGITAL SERVICE BOOKLET

171 260B AIRBAGLABEL-GERMAN/ENGLISH

171 261B AIRBAGLABEL-FRENCH/ENGLISH

171 262B AIRBAGLABLE-SPANISH/ENGLISH

171 291L FACTORY SALES

171 292L FACTORY SALES

171 295L FACTORY SALES

171 299E RHE LOAD CALC. (RIDE HANDLING ELASTOKINEMATICS)

171 2XXL FEDERAL REPUBLIC OF GERMANY

171 300E CERTIFICATION WEIGHT

171 331B INSTRUMENT CLUSTER LANGUAGE - DUTCH

171 332B INSTRUMENT CLUSTER LANGUAGE - GERMAN

171 335B INSTRUMENT CLUSTER LANGUAGE - BRITISH ENGLISH

171 336B INSTRUMENT CLUSTER LANGUAGE - ITALIAN

171 337B INSTRUMENT CLUSTER LANGUAGE - FRENCH

171 338B INSTRUMENT CLUSTER LANGUAGE - SPANISH

171 339B INSTRUMENT CLUSTER LANGUAGE - PORTUGUESE

171 342B INSTRUMENT CLUSTER LANGUAGE - AMERICAN ENGLISH

171 345U BRIGHT BLUE - METALLIC PAINT

171 348B INSTRUMENT CLUSTER LANGUAGE - JAPANESE KATAKANA

171 352U LINARITE BLUE - METALLIC PAINT

171 359U TANZANITE BLUE METALLIC FINISH

171 362L LUEG, BOCHUM

171 372U LAZULITHE BLUE - METALLIC PAINT

171 401L NATO SALES

171 402L TOURISTS (WITHOUT USA)

171 403L FOREIGN DIPLOMATS

171 406L DIRECT BUSINESS CENTER INLAND/EXPORT

171 407L AMG DIRECT BUSINESS

171 408K PAINTWORK, ORANGE

171 409K RADIO LOW END EXPORT

171 410K RADIOHIGH-END

171 411K RADIOHIGH-END EXPORT

171 412K RADIO LOW-END SINGLE-CD

171 412L INTERNATIONAL CONTRACT PARTNERS

171 413L INTERNATIONALAUXILIARYORGANIZATION

171 414K TELEMATIC SAFETY PACKAGE

171 416K TELEMATIC INFOSERVICE

171 417K TELEMATIC SAFETY SYSTEM

171 430U PREHNITE GREEN METALLIC FINISH

171 470K AUTOMATIC TRANSMISSION 7-SPEED

171 471K HEAD UNIT ENTRY

171 472K ASSISTANT STEERING

171 473K 16" LIGHT ALLOY WHEEL ALL ROUND VAR. 1

171 474K LA 16" WHEEL WITH MIXED TIRES

171 475K LA 17" WHEEL, MIXED TIRES (VAR. 1)

171 476K ADVANCED LANE DEPARTURE WARNING (ALDW)

171 489K REPL. W246 F.PASS.SEAT BY W169 F.PASS.SEAT ADAPTED

171 490K INSTALLATION W169 DRIVER SEAT IN INITIAL BODY W246

171 491K BLUE TOOTH (FIXED WITH INTEG. PORTABLE CTEL CON.)

171 492K REMOTE SOFT-TOP ACTUATION

171 496K 17" LIGHT ALLOY WHEEL ALL-ROUND VAR. 4

171 497K WHEEL 7.5X17/8.5X17 225-45/245-40 R17 (VAR.2)

171 498K WHEEL 7.5X17/8.5X17 225-45/245-40 R17 (VAR.3)

171 499K WHEEL 7.5X18/8.5X18 225-45/245-35 R18

171 9XXL AUSTRALIA / PACIFIC

171 B08 HORNS FOR ASIA

171 FR ROADSTER

171 GA AUTOMATIC TRANSMISSION

171 GM MANUAL TRANSMISSION

171 H12 WOOD FINISH, LIGHT BURRED WALNUT VENEER

171 H73 CARBON HIGH-GLOSS TRIM

171 H74 TRIM PARTS - CARBON APPEARANCE

171 H80 PLASTIC TRIM PARTS

171 H81 TRIM PARTS - PLASTIC IN SILVER FABRIC DESIGN

171 H86 LEATHER TRIM PARTS

171 K20 AUTOMATIC TRANSMISSION NOISE OPTIMIZATION

171 L LEFT-HAND STEERING

171 M001 ENGINES WITH SUPERCHARGER

171 M010 EVO/VEHICLEUPGRADEENGINES

171 M011 ENGINE WITH PORT INJECTION

171 M271 R4-GASOLINE ENGINE M271

171 M272 V6-GASOLINE ENGINE M272

171 M273 V8-GASOLINE ENGINE M273

171 M30 DISPLACEMENT 3.0 LITER

171 M35 DISPLACEMENT 3.5 LITER

171 M55 CAPACITY 5.5 LITRE

171 NA0 M271 REPO ENGINE WITH OPT. ENGINE COOLING SYSTEM

171 O03 SEAT BELT REMINDER WARNING SERVICE REDUCED

171 O05 TELEPHONE RETROFITTING USA

171 P04 SERVICE CONTRACT 12 MONTH WARRANTY

171 P05 SERVICE CONTRACT 24 MONTH WARRANTY

171 P16 WORLD CUP BLACK BASIC

171 P23 PROMOTIONAL MODEL

171 P25 SPECIAL MODEL 2LOOK EDITION

171 P30 AMG PERFORMANCE PACKAGE

171 P38 SPECIAL MODEL "EDITION 10"

171 P41 FINAL EDITION SPECIAL MODEL

171 P49 MIRRORS PACKAGE

171 P54 THEFT PROTECTION PACKAGE

171 P62 ACTION MODEL (BR169/209/211/171)

171 P64 MEMORY PACKAGE

171 P84 SPORT EDITION

171 P92 COMFORT PACKAGE WITH NAVIGATION "APS"

171 P94 APPEARANCE PACKAGE

171 P98 SPECIAL MODEL "AMG BLACK SERIES"

171 R RIGHT-HAND STEERING

171 R01 SUMMER TIRES

171 R19 LA 16" 5 SPOKE DESIGN WITH MIXED TIRES

171 R35 LA 10-SPOKE DESIGN 17" WITH TIRE MIX

171 R52 LIGHT ALLOY 9-DOPPLESPOKE DESIGN 17" WITH TIRE MIX

171 R66 TIRE WITH RUN-FLAT PROPERTIES

171 R70 18" 5-TWIN-SPOKE LIGHT ALLOY DESIGN WH. W. MIX. T

171 R87 LIGHT-ALLOY-WHEEL 6-DOUBLESPOKE DESIGN 18"TIRE MIX

171 R88 LIGHT-ALLOY-WHEEL 6-DOUBLESPOKE DESIGN 17"TIRE MIX

171 R93 LT. ALLOY 10-SPOKE DESIGN 17" HIGH-GLOSS MIXED T.

171 R95 LIGHT ALLOY 9-DOPPLESPOKE DESIGN 17" WITH TIRE MIX

171 T01 TELEMATICS PACKAGE 01

171 T02 TELEMATICS PACKAGE 02

171 U06 EXTERNAL MIRROR FOR INDIA

171 U07 CLOTHES CARRIER

171 U12 FLOOR MATS, VELOURS

171 U17 ELIMINATION RED SEATBELT

171 U18 AUTOMATIC CHILD SEAT RECOGNITION (AKSE)

171 U22 LUMBAR SUPPORT ADJUSTMENT

171 U23 REAR BAG

171 U26 AMG FLOOR MATS

171 U48 INJECTION VALVES WITH DBW CAPS

171 U71 DVD PLAYER WITH REGIONAL CODE 1, USA

171 U72 DVD PLAYER WITH REGIONAL CODE 2, EUROPE, JAPAN

171 U73 DVD PLAYER WITH REGIONAL CODE 3, TAIWAN, KOREA

171 U74 DVD PLAYER WITH REGIONAL CODE 4, MEXICO, SOUTH AMERICA

171 U75 DVD PLAYER W/ REG. CODE 5, RUSSIA, EASTERN EUROPE

171 U76 DVD PLAYER WITH REGIONAL CODE 6, CHINA

171 W16 DESIGNO TAMO WOOD FINISH DARK/ANTHRACITE

171 W17 DESIGNO BROWN WOOD FINISH

171 W61 TRIM PARTS PAPPEL DESIGNO-NATURAL

171 W80 LEATHER-CLAD TRIM IN DESIGNO LEATHER

171 X00 LEATHER, DESIGNO, UNI-COLOR

171 X11 LEATHER, DESIGNO-BLACK, UNI-COLOR

171 X16 DESIGNO MAURITIUS BLUE SINGLE-COLOR LEATHER

171 X20 ALCANTARA DESIGNO BLACK SINGLE-COLOR LEATHER

171 X21 ALCANTARA DESIGNO ALPACA GRAY SINGLE-COLOR LEATHER

171 X50 LEATHER, DESIGNO, TWO-COLOR

171 X51 TWO-COLOR RED PEPPER DESIGNO LEATHER

171 X52 TWO-COLOR CURCUMA DESIGNO LEATHER

171 X68 LEATHER DESIGNO-SILVER TWO COLOR

171 X70 LEATHER DESIGNO-MYSTIC BLUE TWO COLOR

171 X77 DESIGNO SAND TWO-TONE LEATHER

171 X78 DESIGNO PASTEL YELLOW TWO-TONE LEATHER

171 X79 DESIGNO MYSTIC RED TWO-TONE LEATHER

171 X80 DESIGNO GRAPHITE GREEN TWO-COLOR LEATHER

171 X82 DESIGNO GRAY VIOLET TWO-TONE LEATHER

171 X83 DESIGNO RED TWO-TONE LEATHER

171 X84 LEATHER DESIGNO-CHABLIS ONE COLOUR

171 X86 DESIGNO MAURITIUS BLUE TWO-TONE LEATHER

171 Y83 HEADLINER, DESIGNO ALCANTARA, ANTHRACITE

171 Y85 HEADLINER, DESIGNO ALCANTARA, ALPACA GRAY

171 Y90 DESIGNOBLACK FABRIC HEADLINING/WINDOW FRAME

171 Y94 SELECTOR LEVER IN LEATHER-/DESIGNO-WOOD VERSION

171 Y95 STEERING WHEEL IN LEATHER-/DESIGNO-WOOD DESIGN

171 Y96 2-TONE DES. LEATHER ST. WH.+DES. LEATHER SEL. LEV.

171 Y97 TRIM PARTS, IN DESIGNO-COLOR, LEATHER-COVERED

171 Y98 ROLL-OVER BAR, IN DESIGNO-COLOR, LEATHER-COVERED

SLK200 in designo havana metallic (code 042)

186

Color and interior choices

The following chart shows the non-designo colors that were available for the R171 owner around 2008 in the US. During the car's lifecycle no all of them were available at all times. In 2011 for example, of the nine metallic color options only seven were still offered. Most of these paints were available also in other markets, but occasionally Daimler-Benz decided to offer them under different names. "Tansanite Blue" was for example in other markets offered as "Capri Blue" or "Calcite White" as "Arctic White". "Prehnite Green" was offered in 2005 only on special request and dropped after 2009.

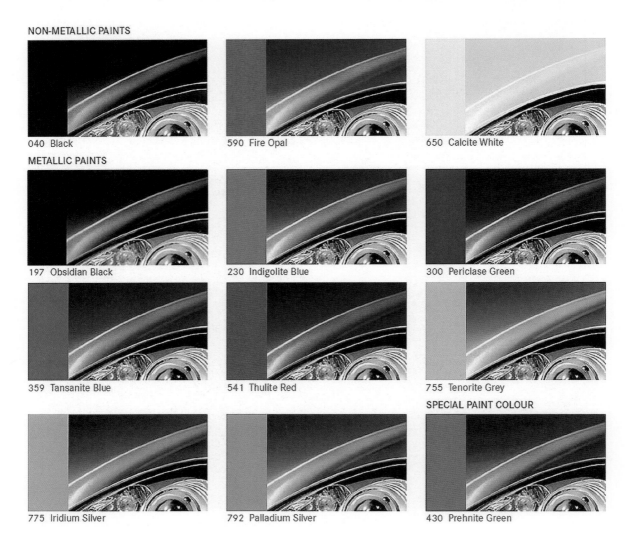

NON-METALLIC PAINTS

040 Black

590 Fire Opal

650 Calcite White

METALLIC PAINTS

197 Obsidian Black

230 Indigolite Blue

300 Periclase Green

359 Tansanite Blue

541 Thulite Red

755 Tenorite Grey

SPECIAL PAINT COLOUR

775 Iridium Silver

792 Palladium Silver

430 Prehnite Green

188

The picture below shows the non-designo interior options that were available. Except "Natural Beige", which was only available in nappa-quality, the other colors were the same for standard leather and nappa leather. Due to the number of possible combinations, designo options are not shown here.

Leather upholstery was offered in Europe only as option. Standard upholstery was black fabric 081. Nappa leather codes had a "5" in the middle of the code, standard leather in the same color came with a "0" or "1" instead. That way standard alpaca grey, also called *ash*, changed from 808 to 858, when chosen as nappa leather.

Standard trim on the center console was aluminum (Avus Silver"), wood applications were optional. The SLK55 came standard with carbon fiber trim.

INTERIOR UPHOLSTERY

808

Alpaca Grey

815

Orient Beige

801

Black

874

Natural Beige
(nappa only)

817

Gullwing Red

TRIM

H80 Avus Silver

H12 Light Burr Walnut

736 Black Ash

H74 Carbon effect

designo TRIM

W61 Natural Poplar

Technical specifications

GENERAL DATA:

SLK200 Kompr., SLK280 / 300, SLK350, SLK55 AMG
(R 171 E18 ML – E55)

Years of Manufacture: March 2004 – March 2011

Price at Introduction:

SLK200 K:	33,520.- € (2004)
SLK200 K:	36,500.- € (2008)
SLK280/300:	39,400.- € (2004)
SLK280/300:	41,860.- € (2008)
SLK350:	43,380.- € (2004)
SLK350:	46,980.- € (2008)
SLK55 AMG:	69,830.- € (2004)
SLK55 AMG:	69,050.- € (2008)
SLK55 AMG Black Series:	107,300.- € (2006)

Chassis/Body: Steel Unit Body

Exterior Dimensions:

Total Length:	160.7 inches (4,082 mm)
Total Width:	70.4 inches (1,777 mm)
Height at curb weight:	51 inches (1,296 mm)
Wheelbase:	95.7 inches (2,430 mm)
Front Track:	60.2 inches (1,530 mm)
Rear Track:	60.7 inches (1,541 mm)

Curb Weight:

SLK200 K:	2,899 lb (1,315 kg)
SLK280/300:	3,009 lb (1,365 kg)
SLK350:	3,064 lb (1,390 kg)
SLK55 AMG:	3,230 lb (1,465 kg)
SLK55 AMG Black Series:	3,131 lb (1,420 kg)

Fuel Tank:

All models:	18.5 gal. (70 liters)

Turning Circle:
34.6 ft (10.55 m)

Drag Coefficient:
0.32 (SLK200 K)

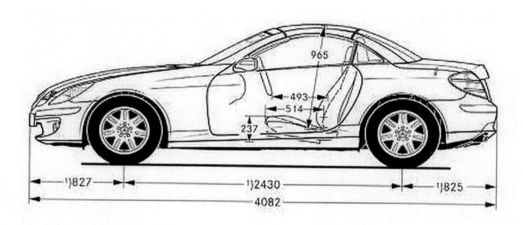

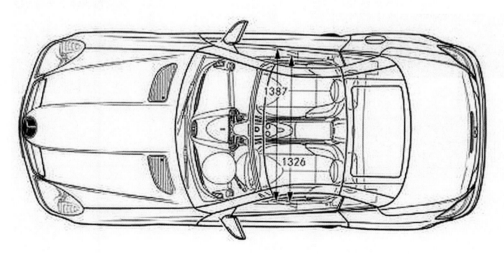

ENGINES:

SLK200 K, 171.442, '04:	4 Cylinder M271 E18 ML, 271.944
Capacity:	109.8 cu inches (1,796 cc)
Configuration:	Front mounted, longitudinal, inline
Head:	Pushrod and rocker actuated, DOHC,
	Variable valve timing
Fuel System:	Indirect injection
Bore and Stroke:	3.23 x 3.35 inches (82 x 85 mm)
Aspiration:	Roots Eaton supercharger
Power:	163 DIN hp @ 5.500 rpm
Torque:	240 Nm @ 3.000 rpm (177 ft/lb)
Compression Ratio:	1:9.5

The SLK200 was the most popular R171

SLK200 K, 171.445, '08:

	4 Cylinder M271 E18 ML, 271.954
Capacity:	109.8 cu inches (1,796 cc)
Configuration:	Front mounted, longitudinal, inline
Head:	Pushrod and rocker actuated, DOHC, Variable valve timing
Fuel System:	Indirect injection
Bore and Stroke:	3.23 x 3.35 inches (82 x 85 mm)
Aspiration:	Roots Eaton supercharger
Power:	184 DIN hp @ 5.500 rpm
Torque:	250 Nm @ 2.800 rpm (184 ft/lb)
Compression Ratio:	1:9.5

SLK280/300, 171.454:

	V6 Cylinder M272 E30, 272.942
Capacity:	182.2 cu inches (2,996 cc)
Configuration:	Front mounted, longitudinal, inline
Head:	Pushrod and rocker actuated, DOHC
Fuel System:	Bosch HFM Multipoint injection
Bore and Stroke:	3.46 x 3.23 inches (88 x 82.1 mm)
Aspiration:	Natural
Power:	231 DIN hp @ 6.000 rpm
Torque:	300 Nm @ 3.500 rpm (221 ft/lb)
Compression Ratio:	1:11,1

SLK350 K, 171.456, '04:

	V6 Cylinder M272 E35, 272.963
Capacity:	214.0 cu inches (3,498 cc)
Configuration:	Front mounted, longitudinal, inline
Head:	Pushrod and rocker actuated, DOHC
Fuel System:	Bosch HFM Multipoint injection
Bore and Stroke:	3.66 x 3.39 inches (92.9 x 86 mm)
Aspiration:	Natural
Power:	272 DIN hp @ 6.000 rpm
Torque:	350 Nm @ 2.400 rpm (258 ft/lb)
Compression Ratio:	1:10.7

SLK350 K, 171.458, '08:	V6 Cylinder M272 E35, 272.969
Capacity:	214.0 cu inches (3,498 cc)
Configuration:	Front mounted, longitudinal, inline
Head:	Pushrod and rocker actuated, DOHC
Fuel System:	Bosch HFM Multipoint injection
Bore and Stroke:	3.66 x 3.39 inches (92.9 x 86 mm)
Aspiration:	Natural
Power:	305 DIN hp @ 6.500 rpm
Torque:	360 Nm @ 2.400 rpm (265 ft/lb)
Compression Ratio:	1:11.7

If you wanted performance without AMG costs, the SLK350 was the right answer

SLK55 AMG, 171.473:	V8 Cylinder M113 E55, 113.989
Capacity:	331.9 cu inches (5,439 cc)
Configuration:	Front mounted, longitudinal, inline
Head:	Pushrod and rocker actuated, SOHC
Fuel System:	Bosch HFM Multipoint injection
Bore and Stroke:	3.82 x 3.62 inches (97 x 92 mm)
Aspiration:	Natural
Power:	360 DIN hp @ 5.750 rpm
Torque:	510 Nm @ 2.700 rpm (376 ft/lb)
Compression Ratio:	1:11
SLK55 AMG BS, 171.479:	V8 Cylinder M113 E55, 113.989
Capacity:	331.9 cu inches (5,439 cc)
Configuration:	Front mounted, longitudinal, inline
Head:	Pushrod and rocker actuated, SOHC
Fuel System:	Bosch HFM Multipoint injection
Bore and Stroke:	3.82 x 3.62 inches (97 x 92 mm)
Aspiration:	Natural
Power:	400 DIN hp @ 5.750 rpm
Torque:	520 Nm @ 3.750 rpm (383 ft/lb)
Compression Ratio:	1:11

PERFORMANCE:

0-62 mph (0-100 km/h):

SLK200 K (2004):	7.9 seconds
SLK200 K (2008):	7.6 seconds
SLK280/300:	6.3 seconds
SLK350 (2004):	5.6 seconds
SLK350 (2008):	5.4 seconds
SLK55 AMG:	4.9 seconds
SLK55 AMG Black Series:	4.5 seconds

Maximum speed:

SLK200 K (2004):	230 km/h (143 mph)
SLK200 K (2008):	236 km/h (147 mph)
SLK280/300:	250 km/h (155 mph), el. ltd.
SLK350 (2004):	250 km/h (155 mph), el. ltd.
SLK350 (2008):	250 km/h (155 mph), el. ltd.
SLK55 AMG:	250 km/h (155 mph), el. ltd.
SLK55 AMG Black Series:	280 km/h (174 mph), el. ltd.

Fuel consumption:

SLK200 K (2004), man.:	8.8 l/100km (26.6 mpg US, 32.0 mpg imp.)
SLK200 K (2008), man.:	8.6 l/100km (27.3 mpg US, 32.8 mpg imp.)
SLK280/300 ('05), auto:	11.4 l/100km (20.6 mpg US, 24.7 mpg imp.)
SLK280/300 ('08), auto:	11.1 l/100km (21.1 mpg US, 25.3 mpg imp.)
SLK350 (2004), auto.:	13.0 l/100km (18.1 mpg US, 21.8 mpg imp.)
SLK350 (2008), auto.:	12.2 l/100km (19.2 mpg US, 23.0 mpg imp.)
SLK55 AMG:	17.1 l/100km (13.8 mpg US, 16.7 mpg imp.)
SLK55 AMG BS:	17.9 l/100km (13.1 mpg US, 15.8 mpg imp.)

DRAG TIMES:

0 – ¼ mile:

SLK200 K (2004), man.:	15.7 seconds
SLK200 K (2008), man.:	15.3 seconds
SLK280/300 ('05), auto:	14.5 seconds
SLK280/300 ('08), auto:	14.2 seconds
SLK350 (2004), auto.:	13.9 seconds
SLK350 (2008), auto.:	13.5 seconds
SLK55 AMG:	13.0 seconds
SLK55 AMG Black Series:	12.6 seconds

Speed at ¼ mile:

SLK200 K (2004), man.:	141 km/h / 88 mph
SLK200 K (2008), man.:	146 km/h / 91 mph
SLK280/300 ('05), auto:	154 km/h / 96 mph
SLK280/300 ('08), auto:	156 km/h / 97 mph
SLK350 (2004), auto.:	162 km/h / 101 mph
SLK350 (2008), auto.:	168 km/h / 104 mph
SLK55 AMG:	175 km/h / 109 mph
SLK55 AMG Black Series:	180 km/h / 112 mph

0 – 1km:

SLK200 K (2004), man.:	28.8 seconds
SLK200 K (2008), man.:	27.9 seconds
SLK280/300 ('05), auto:	26.5 seconds
SLK280/300 ('08), auto:	26.0 seconds
SLK350 (2004), auto.:	24.9 seconds
SLK350 (2008), auto.:	24.5 seconds
SLK55 AMG:	23.5 seconds
SLK55 AMG Black Series:	22.8 seconds

Power to weight ratio:

SLK200 K (2004), man.:	91.3 watt/kg (41.4 watt/lb)
SLK200 K (2008), man.:	102.7 watt/kg (46.6 watt/lb)
SLK280/300 ('05), auto:	122.7 watt/kg (55.7 watt/lb)
SLK280/300 ('08), auto:	124.5 watt/kg (56.5 watt/lb)
SLK350 (2004), auto.:	141.8 watt/kg (64.3 watt/lb)
SLK350 (2008), auto.:	158.9 watt/kg (72.1 watt/lb)
SLK55 AMG:	180.9 watt/kg (82.1 watt/lb)
SLK55 AMG Black Series:	207.0 watt/kg (93.9 watt/lb)

Weight to power ratio:

SLK200 K (2004), man.:	11 kg/kW, 8.1 kg/hp, 18.0 lbs/hp
SLK200 K (2008), man.:	9.7 kg/kW, 7.1 kg/hp, 16.0 lbs/hp
SLK280/300 ('05), auto:	8.1 kg/kW, 6.0 kg/hp, 13.4 lbs/hp
SLK280/300 ('08), auto:	8.0 kg/kW, 5.9 kg/hp, 13.2 lbs/hp
SLK350 (2004), auto.:	7.1 kg/kW, 5.2 kg/hp, 11.6 lbs/hp
SLK350 (2008), auto.:	6.3 kg/kW, 4.6 kg/hp, 10.4 lbs/hp
SLK55 AMG:	5.5 kg/kW, 4.1 kg/hp, 9.1 lbs/hp
SLK55 AMG Black Series:	4.8 kg/kW, 3.6 kg/hp, 7.9 lbs/hp

Co2 emissions:

SLK200 K (2004):	209 g/km
SLK200 K (2008):	182 g/km
SLK280/300:	231 g/km
SLK350 (2004):	255 g/km
SLK350 (2008):	227 g/km
SLK55 AMG:	288 g/km
SLK55 AMG Black Series:	293 g/km

TRANSMISSION:

6-speed Manual:	1st: 4.46:1
	2nd: 2.61:1
	3rd: 1.72:1
	4th: 1.24:1
	5th: 1:1
	6th: 0:84
5-speed Automatic:	1st: 3.95:1
	2nd: 2.42:1
	3rd: 1.49:1
	4th: 1.00:1
	5th: 0.83:1
7G Tronic Automatic:	1st: 4.38:1
	2nd: 2.86:1
	3rd: 1.92:1
	4th: 1.37:1
	5th: 1:1
	6th: 0:82
	7th: 0:73
Clutch:	Single cushion disc, dry plate clutch (man.trans.)

Rear axle ratio:

SLK200 K (2004):	3.46 (5-speed automatic: also 3.46)
SLK200 K (2008):	3.27 (5-speed automatic: also 3.27)
SLK280/300:	3.27 (7G Tronic automatic: also 3.27)
SLK350 (2004):	3.27 (7G-Tronic automatic: also 3.27)
SLK350 (2008):	3.07 (7G-Tronic automatic: 3.27)
SLK55 AMG/BS:	3.06

CHASSIS:

Exhaust:	Twin tail-pipe
Suspension:	
Front:	Independent 3 link, coil springs, anti-roll bar
Rear:	5-arm multilink, coil springs, anti-roll bar
Steering:	Rack-and-pinion, 2.9 turns
Brakes:	
SLK200 K:	
Front:	Vented disc 288 mm (11.3 in)
Rear:	Vented disc 278 mm (11 in)
SLK280/300:	
Front:	Vented disc 300 mm (11.8 in)
Rear:	Solid disc 278 mm (11 in)
SLK350:	
Front:	Vented disc 330 mm (13 in)
Rear:	Solid disc 280 mm (11.4 in)
SLK55 AMG:	
Front:	Vented disc 340 mm (13.4 in)
Rear:	Vented disc 330 mm (13.0 in)
SLK55 AMG Black S.:	
Front:	Vented disc 360 mm (14.2 in)
Rear:	Vented disc 330 mm (13.0 in)
Tires:	
SLK200 K:	205/55 R16 W (front and rear)
SLK280/300:	205/55 R16 W (F), 225/50 R16 W (R)
SLK280/300:	225/45 ZR17 (F), 245/40 ZR17 (R)
AMG standard:	225/40 ZR18 (F), 245/35 ZR18 (R), other sizes optional
SLK55 AMG Black Series:	235/35 ZR19 (F), 265/30 ZR19 (R),

Power & torque curves

SLK200 K, 2005, 163 hp@5,500 rpm, 240 NM (177 ft-lbs)@3,000 rpm

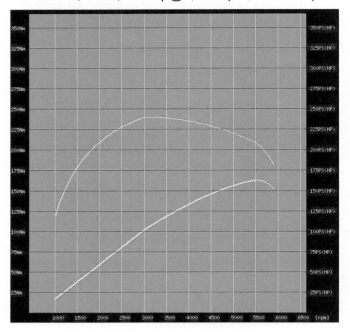

SLK200 K, 2009, 184 hp@5,500 rpm, 250 NM (184 ft-lbs)@2,800 rpm

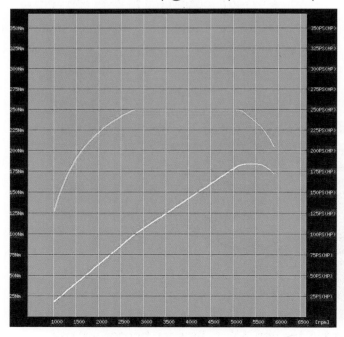

SLK300, 2005, 231 hp@6,000 rpm, 300 NM (221 ft-lbs)@3,500 rpm

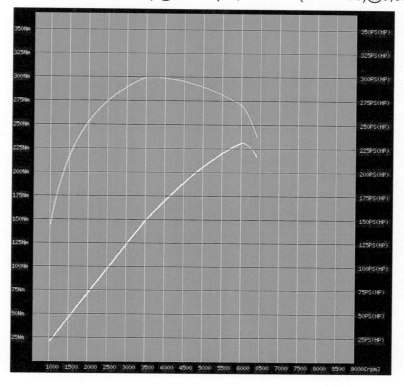

SLK300, 2009, 231 hp@6,000 rpm, 300 NM (221 ft-lbs)@2,500 rpm

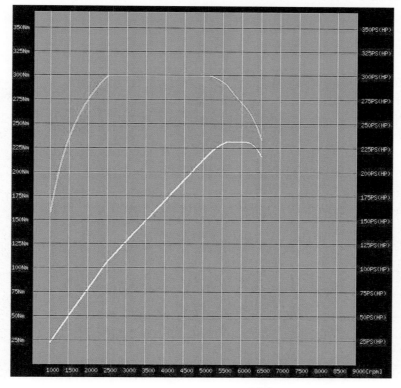

SLK350 , 2005, 272 hp@6,000 rpm, 350 NM (258 ft-lbs)@2,400 rpm

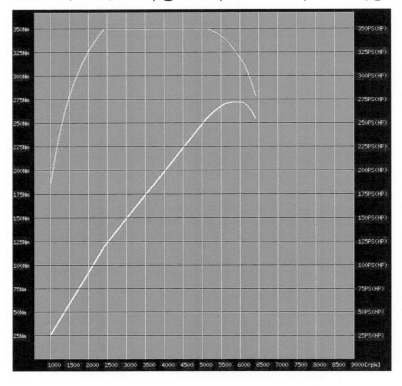

SLK350, 2009, 305 hp@6,500 rpm, 360 NM (265 ft-lbs)@4,000 rpm

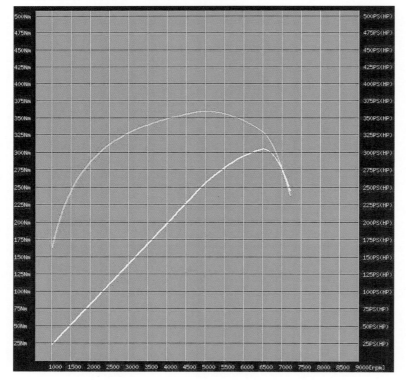

SLK55 AMG, 2005, 360 hp@5,750 rpm, 510 NM (376 ft-lbs)@4,000 rpm

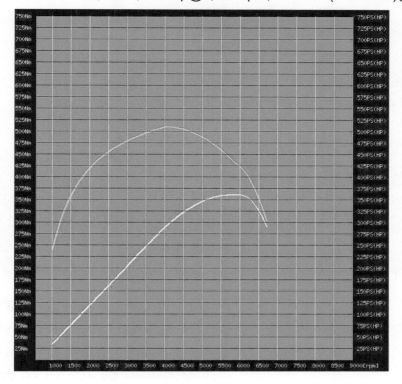

SLK55 AMG Black Series, 2006, 400 hp@5,750 rpm, 520 NM (383 ft-lbs)@3,750 rpm

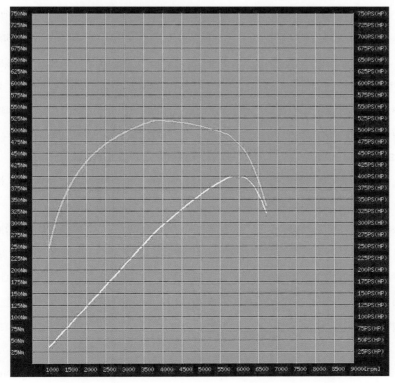

Production data

Type	Code	2003	2004	2005	2006	2007	2008	Total
SLK200K	171442	391	34.378	31.349	22.273	16.573	1.578	**106.542**
SLK280/300	171454		18	9.408	11.070	8.638	6.237	**35.371**
SLK350	171456	80	17.038	15.892	7.397	4.215	283	**44.905**
SLK55	171473	1	1.036	3.904	1.931	1.282	976	**9.130**
Total		**472**	**52.470**	**60.553**	**42.671**	**30.708**	**9.074**	**195.948**

Type	Code	2007	2008	2009	2010	Total	Grand Total
SLK200K	171445	62	16.383	8.822	8.285	**33.552**	**140.094**
SLK280/300	171454	above	Above	1.830	3.214	**5.044**	**40.415**
SLK350	171458	89	4.396	1.211	1.533	**7.229**	**16.359**
SLK55	171473	above	above	227	184	**411**	**196.359**
Total		**151**	**20.779**	**12.090**	**13.216**	**46.236**	**242.184**

Color	in % all markets	Top 10 export markets		Engine	in % all markets
iridum silver	33	Germany	33	SLK200	58
obsidian black	26	United Kingdom	12	SLK280/300	17
black uni	8	Italy	7	SLK350	22
cubanit silver	6	France	3	SLK55 AMG	4
tellure silver	5	USA	3		
fire opal red	4	China	3		
calcit white	3	Japan	3		
benitoit blue	3	Spain	3		
tansanit blue	2	South Africa	2		
palladium silver	2	Canada	2		
alabaster white	2				
thulit red	1				
amber red	1				
jaspis blue	1				
tenorit grey	1				
andradit green	0,2				
periklas green	0,1				
indigolith blue	0,1				
all silver	46				
all black	34				
all red	6				
all blue	5,1				

Acknowledgements

This book would not have been possible without the invaluable information from the Daimler AG archives and from websites such as SLKworld.com and benzworld.org.

Other principal sources for the book in alphabetical order:

Autobild Jan. 2004, Sept.. 2004, Feb. 2008; automotive.com; automuseum-stuttgart.de; auto motor und sport Dec. 2003, Feb. 2004, April 2004, Sept. 2004, Feb. 2007, April 2008, Aug. 20011; emercedesbenz.com, germancarforum.com, "Mercedes-Benz Automobile, Band 2, 1964 – Heute", H. Hofner, H. Schrader

Own archives

All design-, production-, technical- or test-photos and drawings are courtesy of the Daimler AG archives.

Some photos of the SLK are provided by Daimler AG, tuned car photos are by the respective tuning companies.

Other photos are by Bernd S. Koehling or from his collection

The R172 replaced in March 2011 the R171

Other books by the author

Printed books:

MB, The 1930s, The eight-cylinder cars, Part 1: from the Nürburg to the 770 and the G 4
MB, The 1930s, The eight-cylinder cars, Part 2: from the 500K to the 540K and the prototypes

MB, The 170 series: from the 1936 170V to the 170S Cabriolet
MB, The 1950s, The 300 Series: from the 300 Sedan to the 300Sc Cabriolet
MB, The 1950s: The 190SL (W121) and Max Hoffman
MB, The 1950s: The 300SL (W198) Coupe and Roadster
MB, The 1950s, The Ponton Series: from the 180 Sedan to the 220SE Cabriolet

MB, The 1960s, W111: from the 220b to the 220SE Cabriolet
MB, The 1960s, W110: from the 190c to the 230
MB, The 1960s, W112: from the 300SE to the 300SE Cabriolet
MB, The 1960s, W111/112C: from the 220SE Coupe to the 280SE 3.5 Cabriolet
MB, The 1960s, W100: from the 600 to the 600 Pullman Landaulet
MB, The 1960s, W113: from the 230SL to the 280SL
MB, The 1960s, W108/109 six-cylinder Sedan
MB, The 1960s, W108/109 V8 Sedan

MB, The 1970s, W116 Sedan

MB, The modern SL: The R107 from 1971 – 1989
MB, The modern SL, The R129 from 1989 – 2001
MB, The modern SL: The R230 from 2001 – 2011
MB, The modern SL: The R231 from 2011 – 2020

MB, The SLK R170 from 1996 – 2004
MB, The SLK R172 from 2011 onwards

The following books are the first ones the author has written on the subject Mercedes-Benz. They cover several cars of a given period in a single volume, but are less detailed, when it comes to a particular car

MB, The 1950s Volume 1: from the 170V to the 300Sc Roadster
MB, The 1950s Volume 2: from the 180 Ponton to the 300SL Roadster
MB, The 1960s, Volume 1: from the 190c to the 280SE 3.5 Cabriolet
MB, The 1960s, Volume 2: from the 230SL and 600 to the 300SEL 6.3
MB, The early SL cars: from the 300SL Gullwing to the 280SL Pagoda
MB, The modern SL cars: from the 350SL to the SL65 AMG

E-books:

MB, The 170V and 170S (W136) Sedan, OTP and Cabriolets from 1936 – 1955
MB, The 1950s: The 220 (W187) Sedan, Cabriolets and OTP from 1951 – 1955
MB, The 1950s: The 300 (W186, W189, 1951 – 1962) and 300S, Sc (W188, 1951 – 1958)
MB, The 1950s: The 180, 190 (W120, W121) Ponton Sedan from 1953 – 1962
MB, The 1950s: The 219, 220a, S, SE (W 105, W180, W128) Ponton from 1954 – 1960
MB, The 1950s: The 190SL (W121) from 1955 – 1963 and Max Hoffman
MB, The 1950s: The 300SL (W198) Coupe and Roadster from 1954 – 1963

MB, The 1960s: The 220b, 230S (W111) Sedan from 1959 – 1968
MB, The 1960s: The 190c, 200, 230 (W110) Sedan from 1961 – 1968
MB, The 1960s: The 220, 250, 280, 300SE (W111, W112) Coupe/Cabriolet from 1961 – 1971
MB, The 1960s: The 300SE (W112), 1961 – 1965
MB, The 1960s: The 230, 250, 280SL (W113) Pagoda from 1963 – 1971
MB, The 1960s: The 600 (W100), 1963 – 1981
MB, The 1960s: The 250, 280, 300 (W108, W109) six-cylinder Sedan from 1965 – 1972
MB, The 1960s: The 280, 300 (W108, W109) V8 Sedan from 1967 – 1972

MB, The 1970s: The 280, 300, 350, 450, 6.9 (W116) Sedan from 1972 – 1980

MB, The modern SL: The R107 from 1971 – 1989
MB, The modern SL: The R129 from 1989 – 2001
MB, The modern SL: The R230 from 2001 – 2011
MB, The modern SL: The R231 from 2012 – 2020

MB, The SLK R170 from 1996 – 2004
MB, The SLK R171 from 2004 – 2011
MB, The SLK R172 from 2011 onwards

There is one more thing

You have now reached the last page of this book and I sincerely hope that you have liked reading about the history of the SLK R171. Should you have bought the book through Amazon, you have the opportunity to rate it. Your comments will then appear in the review list of my books. If you believe this book is worth sharing, would you be so kind and take a few seconds to let other Mercedes enthusiasts know about it. Maybe they would be grateful. As a small volume author, I will be for sure, as it will help me to gain a bit of recognition. If you will find the time to write a short review, I would like to offer you a **small gift**. You can choose one of my 22 Mercedes e-books and I will send it to your e-mail address free. Of course, I will never spam you.

Please let me know via e-mail, where you have written your review. Here is my e-mail address: bernd@benz-books.com

This book has been printed by an Amazon affiliate printing house, which means that I have unfortunately no control over its printing quality. Should you have received a copy that does not meet your expectations from a printing point of view, please do both of us a favor and ask Amazon for a free replacement. I would also appreciate, if you would inform me about it, if you have the time. Thank you very much for your kind understanding,

Alles Gute,
Bernd